THE MOST EFFICIENT

ENGINE

(The New Carnot Cycle)

John D. Jacoby

THE MOST EFFICIENT ENGINE
(The New Carnot Cycle)

First Edition - 2014

ISBN-10:   1441464395
ISBN-13:   978-1441464392
LCCN: 2013920569

Published by:

John D. Jacoby
PO Box 3451
Ketchum, ID 83340-3451
208-726-3807
JOHNDJACOBY@HOTMAIL.COM

Printed by:

CreateSpace
North Charleston, SC

I dedicate this book, with love, to my father, Joseph, and mother, Evangeline, my three sisters, Dorothy, Phyllis, and Joan and their families, my former wife, Patsy, our three children, Jennifer, John, and Christy, and their families, all of my relatives and in-laws and all of my friends. They have stood by me through good times and bad times and have understood and tolerated my foibles and fables.

ACKNOWLEDGEMENT

I want to thank my daughter, Jennifer, her husband Richard, my son John, his wife Jennifer, and my daughter, Christy, for reviewing my book and suggesting improvements to make it a better book. I especially want to thank my son, John, for spending so much time on it and doing such a good job!

INTRODUCTION

Commuting by car, daily, to work across the Golden Gate Bridge in stop and go traffic was very frustrating. Commuting by bus was not much better. Looking out of the window at the passing scenery, day after day, became too repetitive. I turned to newspapers, crossword puzzles, and magazines to ease the boredom.

One day, browsing through a magazine, I read an article about General Motors Corporation developing a Stirling cycle engine with potential high efficiency. I was interested in energy saving devices and the article caught my fancy. I decided that I wanted to build efficient economical engines powered by solar energy, geothermal energy, and other heat sources.

My quest began. Quite often, on the way home from work, I stopped at the bookstore to browse and occasionally buy books about heat engines. Almost all of those books described the Carnot cycle as the ideal cycle with the highest possible efficiency and as the goal of heat engine designers. However, most of the books concentrated on the Stirling cycle, not on the Carnot cycle. Various books described the Stirling cycle as being capable of the same efficiency as the Carnot cycle, but to achieve that efficiency, regenerators would be required. A regenerator is a feature that recovers waste heat, which is directed back into the engine to increase the engine's efficiency.

To me, if the Carnot cycle were the ideal cycle, it would seem logical to build an engine based on it. After all, a Stirling cycle engine would require regenerators, which, in a practical engine, would add to its cost and make it impossible to equal the Carnot cycle efficiency. Therefore, building a heat engine based upon the Carnot cycle became my goal.

Eventually, my reading and thought processes brought me to the conclusion that despite the Carnot cycle defining the highest heat engine efficiency, an engine designed to operate on the Carnot cycle would be incapable of producing any power at all. Although various books described the Carnot cycle as the goal of all heat engine designers, to achieve such a goal would be meaningless, since a heat engine would be useless if it did not produce any power.

Those thoughts were the beginning of my work developing my heat engine theory and designing an efficient heat engine, and this book is a by-product of that work. Designing my heat engine required estimates for various parameters. There was no literature to help, so I was on my own. I could not afford to build a heat engine based on trial and error, so I applied for a government grant. Although the grant committee acknowledged that my heat engine would run, they turned down my request for a grant because of practicality based upon their estimate of heat transfer.

I felt very let down and gave up for a while. Time healed my wounds and I took to heart the fact that the government had acknowledged that my heat engine would run. I looked at the positive side and began work anew, going back to basics. I started from scratch. I bought books, which included a translation(Mendoza 1977) of Sadi Carnot's famous treatise entitled *Reflections on the Motive Power of Fire, and on Machines Fitted to Develop that Power*, published in France in 1824. The book included his theory on the well-known Carnot cycle, defining the maximum efficiency of all heat engines. His theory is now included in almost all thermodynamic and physics text books, worldwide. I read other heat engine literature and improved my knowledge.

Eventually, I had some inspired thoughts and developed some equations. Some of them became quite complex and I hit an obstacle, I could go no further. Again, I thought that I had failed. I almost gave up. Amazingly, on my way to dreamland one night, as I pondered my failure, I jumped up. For some unknown reason, in my sleepy mind, I had the thought that I had mistakenly used the wrong algebraic sign in one of my equations, a plus sign instead of a minus sign. Sure enough, I found the wrong sign and the obstacle was removed. I corrected my mistake that night so that I would not wake up the next morning thinking that I had dreamed it all.

The next day, I went on with my work, although I was never sure where that work would lead me. Errors in developing my equations continued to plague me, with wrong signs, typing errors, and other silly mistakes, occurring quite often. I persisted, though, and finished developing and correcting my equations, included in this book.

Having been born into the working class, all of my heat engine work took place in my spare time over a period of many years. In a way, that might have been helpful because I developed new ideas along the way. Now, Interested in the importance of my work, I want to expose it to others who might be better qualified to judge it.

Hoping to introduce my book to a broad spectrum of readers, I have tried to keep it as simple, informative, revealing, and interesting as this subject can be. Here is my book:

The Most Efficient Engine ~ *John D. Jacoby*

TABLE OF CONTENTS

__1953__   I hope you enjoy the book! I welcome any comments to make this a better book. - Sincerely, *John D. Jacoby*

JOHNDJACOBY@HOTMAIL.COM

CHAPTER 1 – A LITTLE HISTORY

Heat engines convert heat energy into mechanical energy to perform useful work.

Amazing feats of construction took place long before heat engines were even a dream. Examples include the Greek Parthenon, the Roman Coliseum, the Pyramids of Egypt, the Great Wall of China, and the Taj Mahal in India, among many others. Thousands of people were said to be involved in their construction.

Human beings (many were slaves) and animals appeared to provide the power required. No trucks, so elephants were used in the construction of the Taj Mahal. Eventually, other sources of natural power were used, such as wind and waterpower.

The eighteenth century saw the first heat engines appearing as useful machines. They were crude and inefficient. There was no science of thermodynamics and the building of heat engines was largely a trial and error process. However, the industrial revolution was beginning. Mathematics (such as analytical geometry and calculus), mechanics, and other sciences were progressing. Crude steam engines were appearing and were being put to use in manufacturing, pumping of water, and supplying power to operate ironworks, smelters, textile mills, etc.

Despite the progress in mathematics and mechanics, little was known about the nature of heat. Until the science of thermodynamics in the nineteenth century, many of the world's respected scientists considered heat as a material substance. Little was known about the relationships between the different forms of energy.

The nineteenth century saw the beginning of thermodynamics. Probably the greatest scientific breakthrough was the contribution of Sadi Carnot to the science of thermodynamics. In a technical paper(Carnot 1824), which he published in 1824, he described the conditions necessary for a heat engine to develop utmost efficiency. His theory described a heat engine cycle having an efficiency, which no heat engine could exceed when operating within the same temperature range. Unfortunately, Sadi Carnot died at the age of thirty-six, in 1832. However, his contributions live on.

In the last half of the eighteenth century, the Industrial Revolution began in Great Britain and spread through much of Europe and the United States. Machines replaced manual labor and transformed agricultural economies into industrial ones. It brought economic improvement for most people in industrialized societies, and with it, greater prosperity and improved health. However, the Industrial revolution was not without negative impacts. It spawned the creation of factory pollutants and vehicle pollutants. It also led to much greater land use, which has harmed the natural environment with extensive loss of habitat for animals and plants.

Carnot, by defining the conditions required for greater efficiency of engines, reduced the impacts by speeding up the development of more efficient engines. Now, we have jet planes which circle the globe and space vehicles which land on other planets.

CHAPTER 2 - SOME THERMODYNAMICS

For the understanding of this book, some knowledge of thermodynamics is desirable. Just the word, thermodynamics, makes the subject seem formidable, but in this chapter you will find thermodynamic concepts that will help with understanding of this book.

It will be useful to refer back to this chapter as you read the rest of this book.

Thermodynamics is the science concerned with the relations between heat and work, and the conversion of one into the other. Conversion of heat into work is the subject of this book.

Temperature is a property of matter. Everyone is aware of hot and cold, hot indicating a higher temperature and cold indicating a lower temperature. However, what is temperature? Temperature is a measure of the molecular activity of matter. As an example of molecular activity, air molecules at atmospheric pressure and at thirty-two degrees Fahrenheit travel about one thousand miles per hour, on average, between collisions with other molecules. If the temperature is raised to five hundred degrees Fahrenheit, the molecules travel about four thousand miles per hour on average. Wow! No wonder five hundred degrees feels hot. The air molecules are pelting you at an average of about four thousand miles per hour. O-U-C-H!

Degrees express temperature using various scales, typically Fahrenheit degrees or Celsius degrees. The freezing point of water, 32 degrees Fahrenheit equals 0 degrees Celsius. The boiling point of water, 212 degrees Fahrenheit is equal to 100 degrees Celsius. Therefore, one degree Fahrenheit is equal to 5/9 of one degree Celsius.

Two other temperature scales are useful and needed in thermodynamics, degrees Rankine and degrees Kelvin. One Rankine degree is equivalent to one Fahrenheit degree. One Kelvin degree is equivalent to one Celsius degree. Zero degrees on the Rankine scale is equal to minus 459.67 degrees Fahrenheit. Zero degrees on the Kelvin scale is equal to minus 273.15 degrees Celsius. Zero degrees Rankine and Zero degrees Kelvin is known as absolute zero, the hypothetical point where all molecular activity ceases.

John D. Jacoby

Heat is energy in transition from one body to another due to a temperature difference. Heat flows from a higher temperature body to a lower temperature body. For example, when you are outside on a cold day, you feel cold because heat is flowing from your warm body to the colder surrounding air. If you then enter a warm room, you feel warmer because heat is flowing from the warmer air to your cold body.

Work is energy of translation resulting from a force moving through a distance. For example, if a one pound weight were lifted from the floor to a three foot high table, three foot-pounds of work would have been performed. If the one pound weight were lifted from the floor to the top of a six foot high refrigerator, six foot-pounds of work would have been performed. If the weight had been two pounds instead of one pound, the work in the first case would have been six foot-pounds, and in the second case, the work would have been twelve foot-pounds.

Power is the rate of performing work. If a one pound weight were lifted from the floor to the three foot high table in one minute, the power used would have been three foot pounds a minute. If the weight had been two pounds, the power used would have been six foot-pounds a minute.

Horsepower is a unit of power equal to thirty-three thousand foot-pounds of work performed in one minute. If a man who weighed one hundred and sixty-five pounds climbed a two hundred foot high hill in one minute, he would have used one horsepower. If he had climbed it in four minutes, he would have used one-fourth horsepower. If he had lifted a thirty three hundred pound horse ten feet in one minute, he would have used one horsepower.

Working fluid is a material substance, such as a gas or a liquid, which absorbs, rejects, or transports energy during a thermodynamic process or cycle. The working fluid in your car's engine is air, which is mixed with gasoline and ignited within a cylinder. The heated air expands and forces the piston down. The working fluid, air, performs work.

A *thermodynamic process* occurs when a working fluid changes from one thermodynamic state to another thermodynamic state by the addition or removal of heat or work or both. In this book, six thermodynamic processes will be used.

An *isothermal expansion* is a thermodynamic process during which a working fluid, such as air, receives heat while expanding at a constant temperature and at a decreasing pressure, while performing work.

An *isothermal compression* is a thermodynamic process during which a working fluid rejects heat, while being compressed at a constant temperature, absorbing work.

An *isentropic expansion* is a thermodynamic process during which a working fluid expands with no heat transfer while performing work, its temperature dropping from a higher temperature to a lower temperature.

An *isentropic compression* is a thermodynamic process during which a working fluid is compressed with no heat transfer while absorbing work, its temperature rising from a lower temperature to a higher temperature.

A *constant volume addition of heat* is a thermodynamic process during which heat is added to a working fluid at constant volume, its pressure increasing from a lower pressure to a higher pressure and its temperature rising from a lower temperature to a higher temperature.

A *constant volume rejection of heat* is a thermodynamic process during which heat is removed from a working fluid at constant volume, its pressure dropping from a higher pressure to a lower pressure and its temperature dropping from a higher temperature to a lower temperature.

A *thermodynamic cycle* is a series of thermodynamic processes, wherein a working fluid undergoes those processes and returns to its original state. Typically, thermodynamic cycles are repeated over and over, as in the case of heat engine cycles. *The Most Efficient Engine* cycle is depicted on a temperature-entropy diagram in Chapter 5.

Entropy is a property of matter. It is a property that is difficult to explain, but it is a measure of the unavailable energy during a thermodynamic process. As an example, a working fluid, undergoing an isothermal expansion, receives heat and converts it to work. The heat energy is used up and is no longer available.

J. Willard Gibbs, an outstanding American physicist has been known for making the temperature-entropy diagram an important tool for representing and analyzing thermodynamic processes(Sandfort 1962). The temperature-entropy diagram is used throughout this book, which might not have been written without it. Thank you, Mr. Gibbs!

The six thermodynamic processes used in this book will be depicted on temperature-entropy diagrams. Those depictions will be used in the graphical presentation of the Carnot cycle, the Stirling cycle and the MEE cycle. From here on, temperature-entropy will be designated as T-S, for convenience.

The Carnot cycle will be represented using the isothermal expansion, isentropic expansion, isothermal compression, and isentropic compression processes. The Stirling cycle will be represented using the isothermal expansion, constant volume expansion, isothermal compression and constant volume compression processes. The MEE cycle will use the same processes as the Carnot cycle.

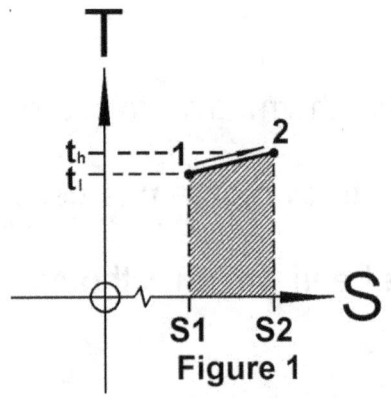

Figure 1

Figure 1 shows a T-S diagram with a generic thermodynamic process depicted thereon. It will provide some important concepts to help with the understanding of this book. The vertical axis, T, is the temperature axis, its positive direction being upward. The horizontal axis, S, is the entropy axis, its positive direction being to the right. Absolute zero of temperature is at the intersection of the two axes. Zero entropy is also at the intersection of the axes. Absolute values of entropy are unknown, so the entropy axis is shown as a broken line. Only relative values of entropy are known.

The generic thermodynamic process, mentioned earlier, is represented by the line from 1 to 2. The small arrow parallel to the line indicates the direction of the process, from 1 to 2. The temperature of the working fluid increases from t_l to t_h. Its entropy increases from S1 to S2. The hatched area below the line 1-2 represents the heat added to the working fluid during the thermody-

namic process. In other words, if a process proceeds from left to

right, it indicates that heat is added to the working fluid and its en-

tropy increases. If the process proceeds in an upward direction,

its temperature increases. So, Figure 1 depicts a process, where-

in the working fluid receives heat, its entropy increases from S1 to

S2, and its temperature increases from t_l

to t_h.

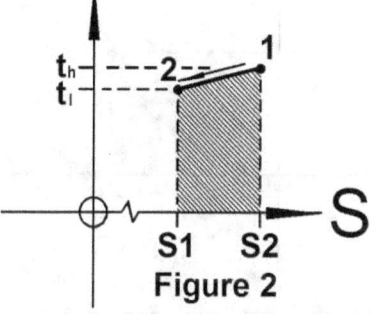
Figure 2

Figure 2 shows a T-S diagram with

another generic thermodynamic process

depicted thereon. Note that it proceeds in the opposite direction

from the process of Figure 1. The small arrow indicates the direc-

tion of the process, from 1 to 2. The temperature of the working

fluid, in this case, drops from t_h to t_l. The entropy decreases from

S2 to S1. The hatched area below the line 1-2 indicates heat re-

moved from the working fluid. Notice that the lines of the hatched area progress from upper left to lower right, just the opposite of Figure 1. So, in this book, T-S diagrams, hatching from upper left to lower right indicates heat removed and upper right to lower left indicates heat added to the working fluid.

Now, specific thermodynamic processes will be described, rather than the two generic processes of Figure 1 and Figure 2.

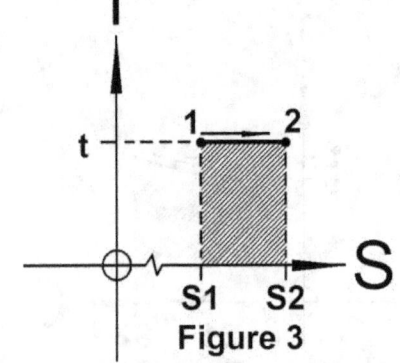

Figure 3

Figure 3 shows an isothermal expansion process depicted upon a T-S diagram. It proceeds from left to right on the diagram, from 1 to 2. Since an isothermal expansion occurs at a constant temperature, it is shown as a horizontal line, the temperature of the working fluid remaining constant at temperature, t. The entropy of the working fluid increases from S1 to S2. The heat added is represented by the shaded area below the line 1-2.

Next, an isothermal compression process will be depicted upon a T-S diagram.

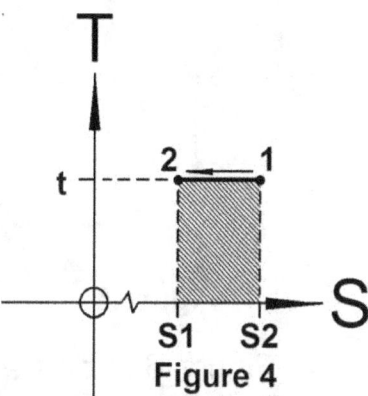

Figure 4

Figure 4 shows an isothermal compression process depicted upon a T-S diagram. It proceeds from right to left on the diagram, from 1 to 2. Since an isothermal compression occurs at constant temperature, it is shown as a horizontal line, the temperature remaining constant at temperature, t. The en-

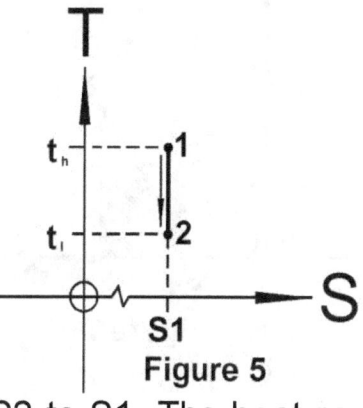

Figure 5

tropy of the working fluid decreases from S2 to S1. The heat removed is represented by the shaded area below the line 1-2.

Figure 5 shows an isentropic expansion process depicted upon a T-S diagram. It proceeds from high to low, from 1 to 2. Since it is isentropic, there is no change in entropy and the temperature drops from t_h to t_l. The entropy is constant at S1. Since

17

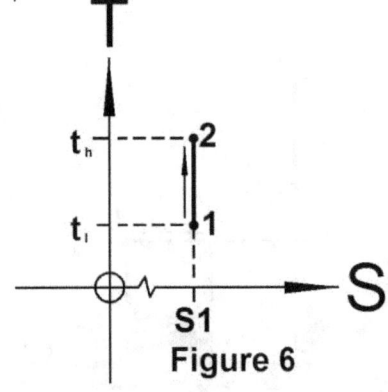

Figure 6

the process is represented by a vertical line, there is no area below the line. No heat is added or removed from the working fluid.

Figure 6 shows an isentropic compression process depicted upon a T-S diagram. It proceeds from low to high, from 1 to 2. Since it is isentropic, there is no change in entropy and the working fluid temperature rises from t_l

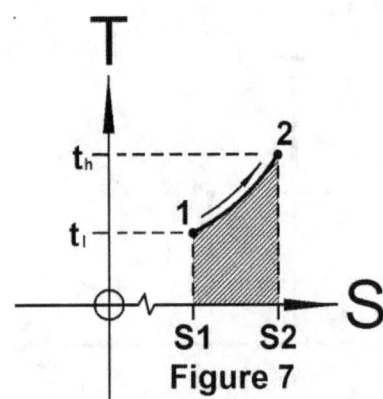

Figure 7

to t_h. The entropy is constant at S1. As in the isentropic expansion of Figure 5, since the process is represented by a vertical line, there is no area below the line. No heat is added or removed from the working fluid.

Figure 7 shows a constant volume addition of heat process depicted upon a T-S diagram. The temperature of the working fluid rises from t_l to t_h. The entropy of the working fluid increases from S1 to S2. The shaded area below the line 1-2 represents the heat added to the working fluid.

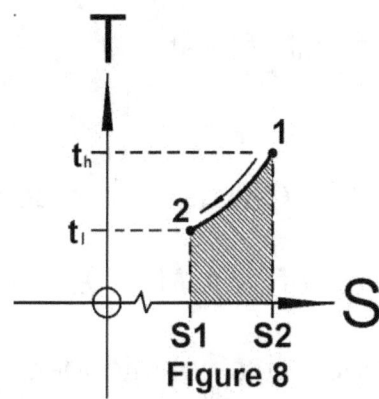

Figure 8

Lastly, Figure 8 shows a constant volume removal of heat process depicted upon a T-S diagram. The temperature of the working fluid drops from t_h to t_l. Its entropy decreases from S2 to S1, and the shaded area below the line 1-2 represents the heat removed from the working fluid.

The processes shown in Figures 3 through 8 will be used in the upcoming chapters as elements of thermodynamic cycles.

In summary, those processes are isothermal expansion, isothermal compression, isentropic expansion, isentropic compres-

19

sion, constant volume addition of heat, and constant volume removal of heat.

Only the thermodynamic principles needed for the presentation of *The Most Efficient Engine*, the MEE, are included in this book. It is not intended to be a course in thermodynamics.

The following two definitions are also included for clarity.

A *heat source* in this book means a source of heat, with a temperature higher than or equal to the temperature of the working fluid, in contact with the working fluid of an engine.

A *heat sink* in this book means a heat receiving body, with a temperature lower than or equal to the temperature of the working fluid, in contact with the working fluid of an engine. In a heat engine, heat is received from a heat source, partially converted to work, and unused heat is rejected to a heat sink.

CHAPTER 3 – THE CARNOT CYCLE

Almost any book of physics or book of thermodynamics has a description of the Carnot cycle.

The Carnot cycle, presented by Sadi Carnot(Carnot 1824), *Reflexions sur la Puissance Motrice Du Feu et Sur Les Machines Propres a Developper Cette Puissance,* defines the maximum possible efficiency of any heat engine operating within given temperature limits. When Carnot published his famous technical paper, entropy and temperature-entropy diagrams were unknown. However, he still correctly postulated his theory. It was not until later in the nineteenth century that entropy and temperature-entropy diagrams became known, giving us tools, which Carnot did not have, to analyze heat engine cycles.

Although Sadi Carnot's famous paper(Carnot 1824) was published in 1824, it did not receive recognition until years later, near the end of the nineteenth century, when classical thermodynamics was established by others including the English physicist, Lord Kelvin, and the German physicist, Rudolf Clausius. Unfortunately, Sadi Carnot died in 1832 at a very young age, but left a lasting legacy.

Now, the Carnot cycle is described using temperature-entropy diagrams, and it will be described herein using such diagrams.

Referring to Figure 9, the Carnot cycle is depicted on a T-S diagram. The cycle consists of four thermodynamic processes.

Beginning at 1, the first process is an isothermal expansion from 1 to 2. Heat is added and the working fluid expands at the constant high temperature, T_H, its pressure reducing and its entropy

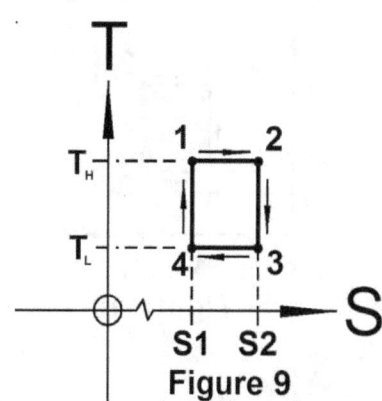

Figure 9

increasing. During the isentropic process, 2-3, the working fluid continues to expand without the addition of heat, its temperature falling from T_H to T_L, its pressure reducing further and its entropy remaining at S2. The cycle continues with process 3-4, an isothermal compression, the working fluid rejecting heat at the constant low temperature, T_L, its pressure increasing and its entropy decreasing from S2 to S1. During the last thermodynamic process, an isentropic compression, the working fluid is compressed with no heat transfer and its temperature increasing from T_L to T_H. Then the cycle begins anew.

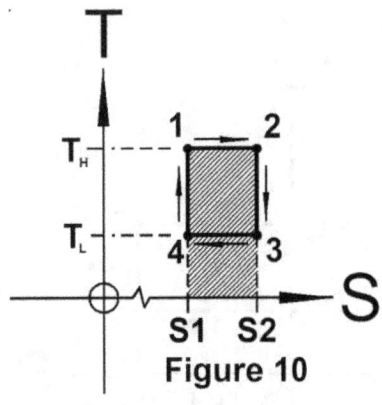

Figure 10

Work is performed during the first two processes and work is absorbed during the last two processes. The work performed exceeds the work absorbed and net work results.

Referring to Figure 10, the shaded area below the isothermal expansion, 1-2 represents the heat added to the working fluid.

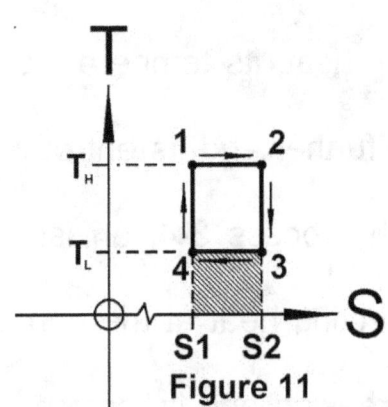

Figure 11

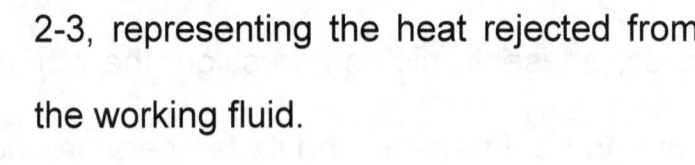

Figure 11 shows the Carnot cycle with the shaded area below the isothermal compression process, 2-3, representing the heat rejected from the working fluid.

Figure 12 shows the Carnot cycle with the shaded portion representing the

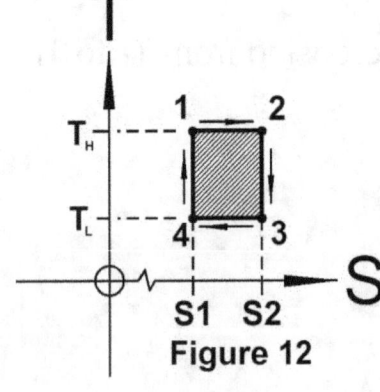

Figure 12

24

difference between the heat added (Figure 10) and the heat re-

jected (Figure 11). The difference is the net heat available to pro-

duce work.

The efficiency of the Carnot cycle engine is defined as the

heat converted to work (the net heat), divided by the heat added.

From FIG. 12, the net heat is the area defined by $(T_H - T_L)(S_2 - S_1)$.

The heat added, referring to FIG. 10, is $T_H(S_2 - S_1)$.

Letting E equal the Carnot cycle efficiency, the follow equa-

tion can be written:

$$E = \frac{(T_H - T_L)(S2 - S1)}{T_H(S2 - S1)}$$
(3.1)

Equation (3.1) can be simplified by canceling the $(S_2 - S_1)$ of

the denominator and the numerator, giving:

(3.2)
$$E = \frac{T_H - T_L}{T_H}$$

Optionally, the equation for efficiency is written as:

(3.3)
$$E = 1 - \frac{T_L}{T_H}$$

Figure 12 is repeated here for ease of reference.

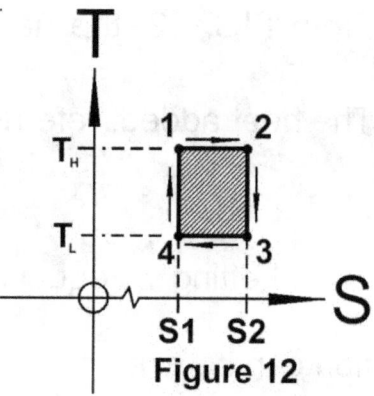

Figure 12

Referring to equation (3.3), the Carnot cycle efficiency can be increased by lowering T_L or by raising T_H, or doing both, thereby decreasing the second term on the right of the equation.

So, in essence, the Carnot theory says:

The efficiency of all heat engines operating within the same temperature limits cannot exceed that of the Carnot cycle using those same temperature limits. Raising the temperature of the heat source and/or lowering the temperature of the heat sink results in an increase of efficiency.

It should be noted that, in reality, no engine can operate on the Carnot cycle, because in order for heat to flow the temperature of the heat source must be at a higher temperature than the working fluid at its high temperature. There must also be a temperature difference between the heat sink and the working fluid. The temperature of the heat sink must be at a lower temperature than the working fluid at its low temperature.

The typical description of the Carnot cycle states that the high temperature of the working fluid is at a temperature infinitesimally lower than the heat source temperature and the low temperature of the working fluid is infinitesimally higher than the temperature of the heat sink. Although this theoretically allows an engine based on the Carnot cycle to run, the power produced would be infinitesimal and meaningless for a real heat engine.

That does not mean that the Carnot cycle is meaningless, far from it. What it means is that the Carnot cycle is not meant to be the ultimate heat engine. It only defines the maximum efficiency for a heat engine operating within a given temperature range.

However, I have read, in the past, that many heat engine designers have considered the Carnot cycle as the ideal goal. An impossible achievement!

CHAPTER 4 - THE STIRLING CYCLE

The Stirling cycle engine has been a popular type of heat engine. Its operation is based upon the theoretical Stirling cycle, which is often touted as capable of achieving the same efficiency as the Carnot cycle. In order to achieve this, regenerators are required, which work by returning waste heat to the engine for reuse by the engine. That should become more clear as you read on.

The Stirling cycle engine was invented by Robert Stirling, a Minister of the Church of Scotland, in 1816. Work on Stirling cycle engines has been carried on by General Motors, Philips Laboratories, and other companies. However, it seems that interest has waned in recent years, possibly because of weight and cost compared to gas and diesel engines, which now have increased efficiencies.

Other research continues, and Stirling engines are used for solar power and space power. Many books have been written about the Stirling cycle engines, although there seems to be little media coverage. However, since the Stirling cycle seemed to be preferred over the Carnot cycle with respect to practical engines, it is probably important to include it in this book.

The Stirling cycle is a theoretical cycle consisting of four thermodynamic processes depicted on a temperature-entropy diagram as shown in Figure 13, with the vertical axis, T, as the temperature axis and the horizontal axis, S, as the entropy axis. The entropy axis is shown broken, since absolute values of entropy are unknown and only relative values are known. The arrows on the axes indicate the positive direction. The smaller arrows parallel to the thermodynamic processes indicate the direction of the processes.

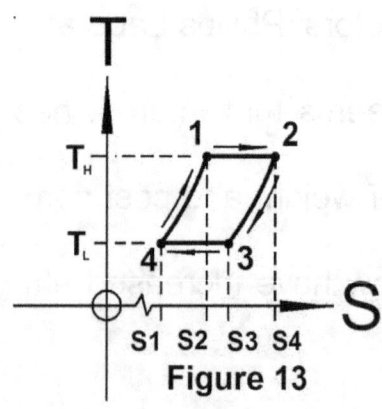

Figure 13

Beginning at 1, the first process, 1-2, is an isothermal expansion with heat being added to the working fluid at the constant high temperature T_H, its pressure reducing and its volume increasing, while performing work. The second process, 2-3, is a constant volume rejection of heat by the working fluid, as its temperature drops from T_H to T_L. No work is performed, since there is no change in volume. The third process, 3-4, is an isothermal compression, with heat being rejected from the working fluid at the constant low temperature, T_L, its pressure increasing and its volume decreasing, while absorbing work. The fourth process, 4-1, is another constant volume process, with the working fluid receiving heat, its pressure rising and its temperature increasing from T_L to T_H. No work is performed, since there is no change in volume. The shaded area of Figure 14, represents the heat added

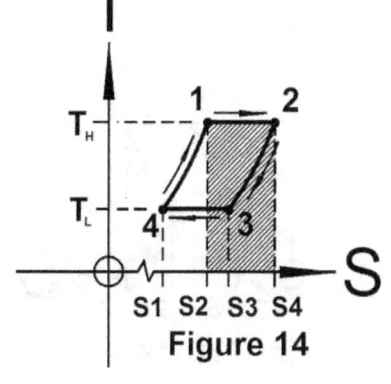

Figure 14

to the working fluid during the isothermal expansion, 1-2.

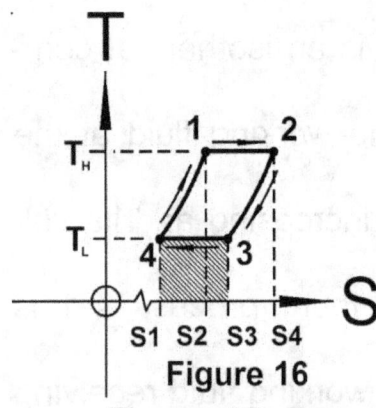
Figure 15

The shaded area of Figure 15 represents the heat rejected from the working fluid during the constant volume process, 2-3.

In Figure 16, the shaded area represents the heat rejected during the isothermal compression, 3-4.

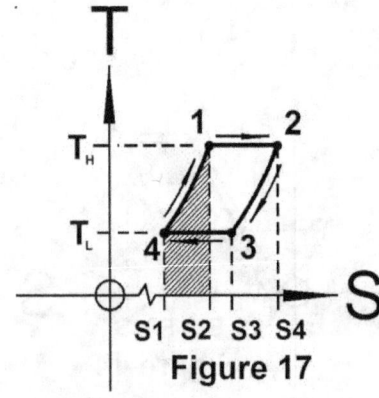
Figure 16

Figure 17 shows a shaded area, representing the heat added during the constant volume process, 4-1.

So, heat has been added during the process, 1-2, and during the process,

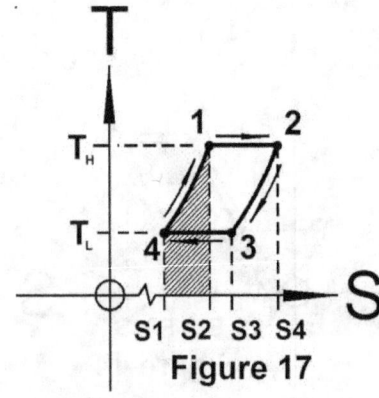
Figure 17

4-1. Heat had been rejected during process 2-3, and during process 3-4.

The difference between the heat added and the heat rejected is the net

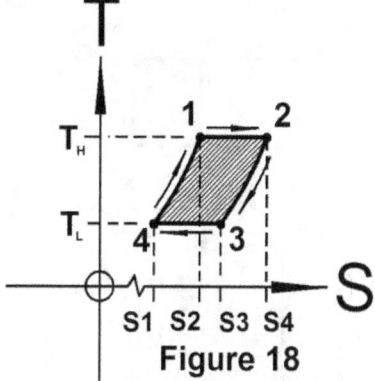

Figure 18

heat, shown by the shaded area of Figure 18. The net heat of the cycle is equivalent to the net work of the cycle.

Now, comparing the Stirling cycle to the Carnot cycle, assume that the isothermal expansion, 1-2, of the Stirling cycle is the same as the isothermal expansion, 1-2, of the Carnot cycle, and the isothermal compression, 3-4, of the Stirling cycle is the same as the isothermal compression, 3-4, of the Carnot cycle. Also, assume that T_H and T_L of both cycles are the same.

The efficiency of both engines is defined as the net heat divided by the heat added.

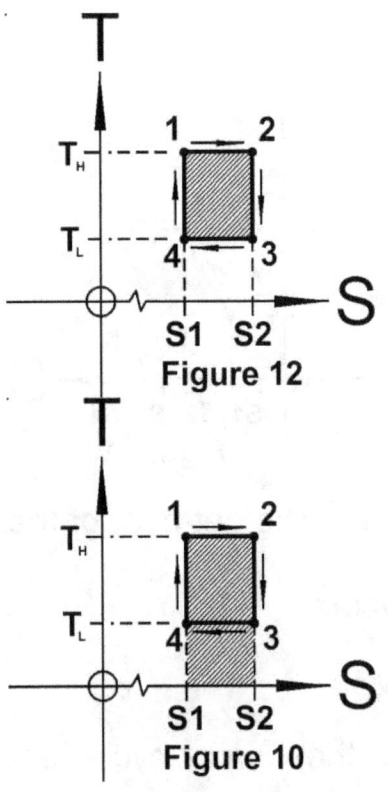

Figure 12

Figure 10

With this simplified analysis, the efficiency of the Carnot cycle can be shown to be greater than that of the Stirling cycle. Figures 10 and 12 of the Carnot cycle and Figures 14, 17, and 18 of the Stirling cycle are repeated here for convenience. Assume that the isothermal processes, 1-2, of Figures 12 and 18, are equal and the temperatures are equal. Therefore, the shaded areas are approximately equal, meaning that the work of the two cycles is approximately equal. The heat added for the Carnot cycle is the shaded area of Figure 10. For the Stirling cycle, the head added is the sum of the shaded areas of Figures 14 and 17. Therefore, the efficiency of the Carnot cycle must be greater than that of the Stirling cycle, since the efficiency of each is the net heat divided by the heat added.

As mentioned earlier, the efficiency of the Stirling cycle can approach that of the Carnot cycle with the use of regenerators. The heat rejected, shown previously in Figure 15, can be recovered and used to supply the heat added, shown in Figure 17. With a one-hundred percent efficient regenerator the efficiency of the Stirling cycle can be made to equal that of the Carnot cycle, but it is not possible to have a one-hundred percent efficient regenerator. So, all of the rejected heat cannot be recovered and reused and the efficiency of the Stirling cycle will, in actuality, be less than the efficiency of the Carnot cycle.

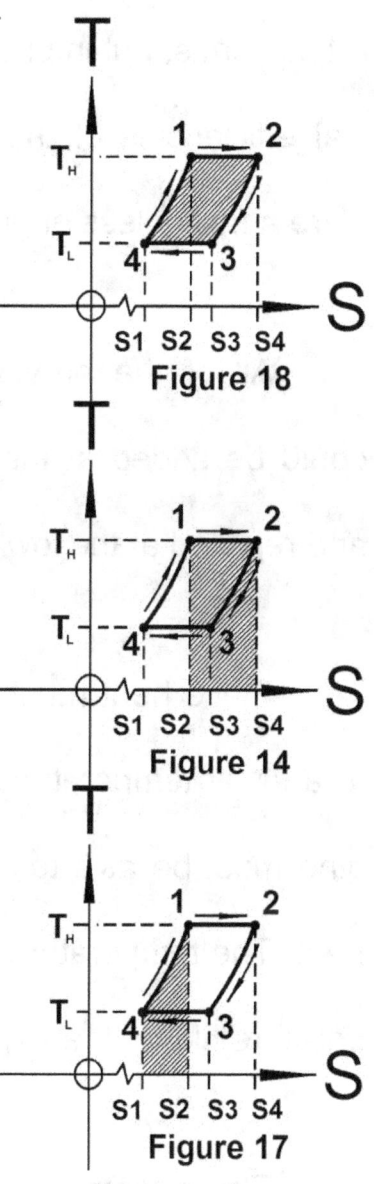

Figure 18

Figure 14

Figure 17

Of course, all of the previous discussions have been in theoretical terms, but in practice it is impossible to achieve the theoretical efficiencies of the Carnot and Stirling cycles. Even if there were a frictionless engine, that efficiency could not be achieved.

All of the previous discussions have assumed that heat could be added at the high temperature, T_H, of the heat source and rejected at the low temperature, T_L, of the heat sink.

Since heat must flow from a higher temperature substance to a lower temperature substance, the working fluid of a heat engine must be at a temperature lower than T_H in order to receive heat. The temperature of the working fluid must also be at a temperature higher than T_L in order to reject heat.

Those temperature differentials must be present for a real heat engine to operate and they will be taken into account in the next chapter, *The Most Efficient Engine Cycle,* which is the heart of this book.

CHAPTER 5 – THE MEE CYCLE

The Most Efficient Engine, hereinafter called the MEE, operates on a cycle, which looks very much like the Carnot cycle when plotted on a temperature-entropy diagram. They both appear as rectangles comprising four thermodynamic processes. One important difference between the Carnot cycle and the MEE cycle is that the MEE cycle is a variable cycle.

The efficiency of the MEE cycle varies from a minimum of zero to a maximum, equal to that of the Carnot cycle. The MEE

37

cycle supplements the Carnot cycle because, as well as defining the maximum efficiency of heat engines operating within the same temperature limits, it gives additional insight on work, power, and efficiency of the engine. The charts and graphs for the MEE cycle will give a good visualization of the parametric relationships.

Chapter 3, *The Carnot Cycle,* and Chapter 4, *The Stirling Cycle* describe well known heat engine cycles. Chapter 5, *The MEE Cycle*, describes a brand new concept for a heat engine cycle.

The MEE should prove very interesting, especially to those who are interested in heat engines. It is mathematically described, using equations, which become complex at times. However, they become complex because of their length. Most of the equations are algebraic equations. Several of them involve the use of calculus. In the end, the charts and graphs present the MEE in a fairly simple manner. If you could get through the previous chapters, you should be able to understand this chapter, although the mathematics might slow you down. Of course, the mathematics is necessary to present the MEE, and you might even enjoy it. So here goes!

As mentioned previously, there must be a difference of temperature between the heat source and the working fluid in order for heat to flow to the working fluid of the engine. There must also be a difference of temperature between the working fluid and the heat sink in order for heat to flow from the working fluid. The MEE cycle is based upon that requirement. Figure 19 depicts the MEE cycle plotted upon a temperature-entropy graph. As before, the vertical axis, T, is the temperature axis and the horizontal axis, S, is the entropy axis. The arrows on the axes indicate the positive direction. The entropy axis is again shown broken, since absolute val-

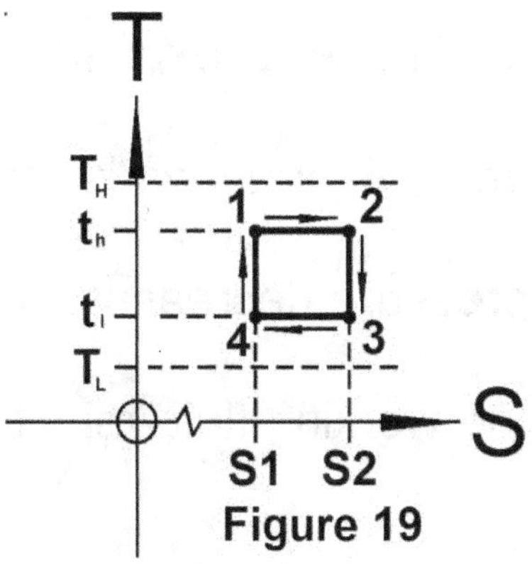

Figure 19

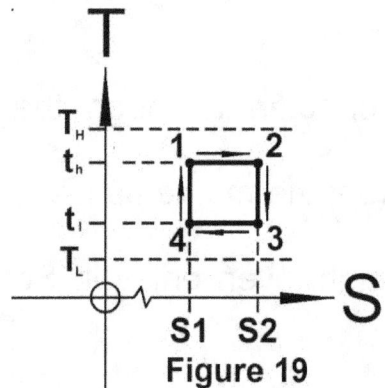

Figure 19

ues of entropy are unknown. Only relative values of entropy can be determined.

T_H is the high temperature of the heat source. T_L is the low temperature of the heat sink. The high temperature of the MEE working fluid is t_h. The low temperature of the working fluid is t_l. S1 is the low entropy and S2 is the high entropy.

Process, 1-2, is an isothermal expansion, the working fluid receiving heat at temperature, t_h, from the heat source at temperature, T_H. The working fluid expands, performs work, and its pressure decreases as it receives heat, its entropy increasing from S1 to S2. Process, 2-3, is an isentropic expansion. The working fluid continues to expand with no heat transfer, at constant entropy, and performs work. Its temperature reduces from t_h to t_l and its pressure decreases. Process, 3-4, is an isothermal compression, the working fluid rejecting heat at temperature, t_l, to the heat sink at temperature, T_L. The working fluid is compressed, its pressure increasing and its entropy decreasing.

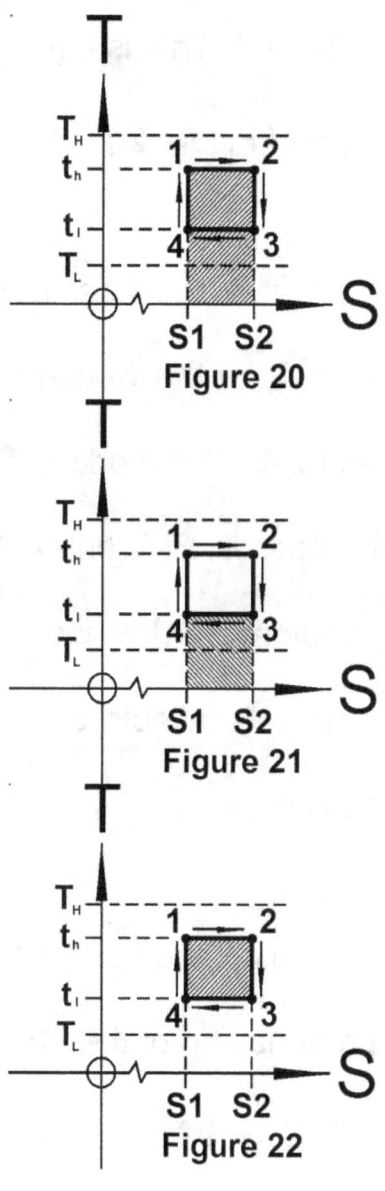

Figure 20

Figure 21

Figure 22

Figure 20 shows the MEE cycle with the shaded area representing the heat added to the working fluid. The heat added is the area t_h times the increase in entropy, S2-S1.

Figure 21 shows the MEE cycle with the shaded area representing the heat rejected from the working fluid. The heat rejected is the area t_l times the decrease in entropy S2-S1.

Figure 22 shows the MEE cycle with the shaded area representing the net heat. The net heat is equal to the heat added minus the heat rejected. It is equal to the area within the rectangle, 1-2-3-4-1. This is equal to $(t_h - t_l)$ times $(s_2 - s_1)$, which is the shaded area of Figure 20 minus the shaded area of Figure 21.

The net heat is equivalent to the net work. Work is performed by the working fluid during the expansion processes, 1-2, and 2-3. The working fluid absorbs work during the compression processes, 3-4, and 4-1. The difference between the work of expansion and the work of compression is the net work. The net work is available to overcome friction and to power an external device.

The previous figures and their descriptions give a visual presentation of the MEE cycle. Equations will be derived in the following pages to express, mathematically, the operation of the MEE.

Charts and graphs will be plotted, using the MEE equations to give another visual presentation of the MEE cycle operation.

Chapter 5 - The MEE Cycle

Following are the terms to be used for the MEE equations:

a = the heat transfer coefficient at the heat source.

b = the heat transfer coefficient at the heat sink.

e = the efficiency of the MEE

E = the Carnot cycle efficiency.

H_A = the heat added to the working fluid during a cycle.

H_R = the heat rejected by the working fluid during a cycle.

H_N = the net heat converted to work during a cycle.

$S1$ = the low entropy of the working fluid.

$S2$ = the high entropy of the working fluid.

t_h = the high temperature of the working fluid.

t_l = the low temperature of the working fluid.

T_H = the heat source temperature.

T_L = the heat sink temperature.

W_N = the net work performed during a cycle.

W_R = the net work relative to the maximum net work.

The heat transfer coefficient, a, and the heat transfer coefficient, b, indicate the heat transfer per degree per cycle to or from the working fluid. The heat transfer coefficients of the heat source and the heat sink are determined by the heat transfer characteristics of the heat source and the heat sink and by the heat transfer areas. If the heat transfer characteristics were identical, then the heat transfer coefficients would be proportional to the areas of the heat source and the heat sink. For example, if the heat transfer coefficients were identical and the area of the heat sink were 1.5 times the area of the heat source, then the heat transfer coefficient, b, would be 1.5 times the heat transfer coefficient, a.

As this chapter progresses, certain Figures will be repeated to allow referral to the Figures without having to thumb through the pages. Remember to refer back to Chapter 2, *Some Thermodynamics*, to refresh your memory if you have forgotten any concepts, which you need to understand as you read further.

Now, the equations!

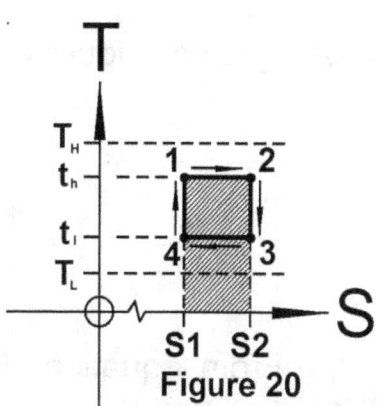

Figure 20

The shaded area of Figure 20 equals the heat added to the MEE cycle during the isothermal process, 1-2. The following equation results:

$$H_A = t_h (S2 - S1) \qquad (5.1)$$

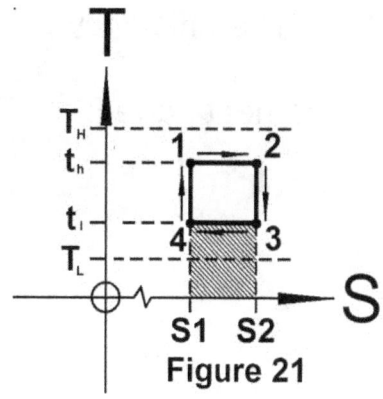

Figure 21

The shaded area of Figure 21 equals the heat rejected from the working fluid of the MEE cycle during the isothermal compression process, 3-4. As above, another equation results for the heat rejected:

$$H_R = t_l (S2 - S1) \qquad (5.2)$$

45

John D. Jacoby

The difference between the heat added (Figure 20) and the
heat rejected (Figure 21) is the net heat: $H_N = t_h(S2-S1) - t_l(S2-S1)$

(5.3) $\qquad\qquad\qquad\qquad H_N = H_A - H_R$

From equation (5.1) and equation (5.2), equation (5.3) is
equivalent to:

(5.4) $\qquad\qquad\qquad H_N = t_h(S2-S1) - t_l(S2-S1)$

The efficiency of the MEE cycle equals the net heat divided
by the heat added:

(5.5) $\qquad\qquad\qquad\qquad e = \dfrac{H_N}{H_A}$

Substituting the right side of equation (5.1) for H_A and the
right side of equation (5.4) for H_N, equation (5.5) becomes:

46

$$e = \frac{t_h\left(S2-S1\right)-t_l\left(S2-S1\right)}{t_h\left(S2-S1\right)} \qquad (5.6)$$

Equation (5.6) can be simplified by canceling $(s_2 - s_1)$ from the numerator and the denominator, giving:

$$e = \frac{t_h - t_l}{t_h} \qquad (5.7)$$

Equation (5.7) gives the efficiency of the MEE in terms of the working fluid temperatures t_h and t_l.

The heat added, as in equation (5.1), is also equal to the heat transfer coefficient, a, times the difference in temperature between the heat source and the working fluid:

$$H_A = a\left(T_H - t_h\right) \qquad (5.8)$$

47

Similarly, the heat rejected, as in equation (5.2), is equal to the heat transfer coefficient, b, multiplied by the difference in temperature between the working fluid and the heat sink:

(5.9)
$$H_R = b(t_I - T_L)$$

The heat added, equation (5.8), minus the heat rejected, equation (5.9), gives the net heat, H_N:

(5.10)
$$H_N = H_A - H_R = a(T_H - t_h) - b(t_I - T_L)$$

Another equation for efficiency can now be written. The efficiency is equal to the net heat divided by the heat added. Using equation (5.8) and equation (5.10), the equation for efficiency is:

(5.11)
$$e = \frac{a(T_H - t_h) - b(t_I - T_L)}{a(T_H - t_h)}$$

Since there are two equations for efficiency now, equation (5.7) and equation (5.11), the right sides of those equations can be set equal to each other, giving:

$$\frac{t_h - t_l}{t_h} = \frac{a(T_H - t_h) - b(t_l - T_L)}{a(T_H - t_h)} \qquad (5.12)$$

Now, the temperature, t_h, can be determined in terms of t_l. Multiplying both sides of equation (5.12) by the product of the denominators gives the following equation:

$$aT_H t_h - at_h^2 - aT_H t_l + at_l t_h = aT_H t_h - at_h^2 - bt_l t_h + bT_L t_h \qquad (5.13)$$

The first two terms on the left side of the equation cancel with the first two terms on the right side of the equation, reducing equation (5.13) to:

$$-aT_H t_l + at_l t_h = bT_L t_h - bt_l t_h \qquad (5.14)$$

Solving for t_h, the following equation results:

(5.15)
$$t_h = \frac{aT_H t_l}{(a+b)t_l - bT_L}$$

Equation (5.15) gives the temperature, t_h, in terms of t_l. Similarly, the temperature, t_l, can be determined in terms of t_h.

Solving equation (5.14) for t_l gives:

(5.16)
$$t_l = \frac{bT_L t_h}{(a+b)t_h - aT_H}$$

Equations (5.15) and (5.16), giving t_h in terms of t_l and giving t_l in terms of t_h, will be useful in some of the subsequent equations to be developed. Now, the net heat can be determined in terms of t_h and, also in terms of t_l. Equation (5.10) can be rewritten by multiplying out the right side of the equation:

(5.17)
$$H_N = aT_H - at_h - bt_l + bT_L$$

Substituting the right side of equation (5.16) for t_l in equation (5.17), the net heat, H_N, can be expressed in terms of t_h. This gives the following equation:

$$H_N = aT_H - at_h - b\left(\frac{bt_hT_L}{(a+b)t_h - aT_H}\right) + bT_L \qquad (5.18)$$

By multiplying out the third term on the right side of equation (5.18), the equation becomes:

$$H_N = aT_H - at_h - \frac{b^2 T_L t_h}{(a+b)t_h - aT_H} + bT_L \qquad (5.19)$$

Equation (5.19) gives the value of net heat, H_N, in terms of the high temperature of the working fluid, t_h.

By finding the derivative of the net heat, H_N, with respect to the high temperature of the working fluid, t_h, the maximum value of the net heat can be determined. Using calculus on Equation (5.19), the derivative is:

(5.20) $\quad \dfrac{dH_N}{dt_h} = 0 - a - \dfrac{\left((a+b)t_h - aT_H\right)\left(b^2 T_L\right) - \left(b^2 T_L t_h\right)(a+b)}{\left((a+b)t_h - aT_H\right)^2} + 0$

To find the temperature, t_h, at which the net heat, H_N, is a maximum, the right side of equation (5.20) is set equal to zero and solved for t_h.

The following equation, results:

(5.21) $\quad a + \dfrac{\left((a+b)t_h - aT_H\right)\left(b^2 T_L\right) - \left(b^2 T_L t_h\right)(a+b)}{\left((a+b)t_h - aT_H\right)^2} = 0$

Now, the equation can be solved for t_h. Multiplying both sides of the equation by the denominator gives:

(5.22) $\quad a\left((a+b)t_h - aT_H\right)^2 + \left((a+b)t_h - aT_H\right)\left(b^2 T_L\right) - \left(b^2 T_L t_h\right)(a+b) = 0$

Squaring, multiplying, and simplifying produce the following equation:

$$a(a+b)^2 t_h^2 - 2a^2(a+b)T_H t_h + a^3 T_H^2 - ab^2 T_H T_L = 0 \quad (5.23)$$

Dividing both sides of equation (5.23) by $a(a+b)^2$ gives:

$$t_h^2 - \left(\frac{2aT_H}{(a+b)}\right)t_h + \left(\frac{a^2 T_H^2 - b^2 T_H T_L}{(a+b)^2}\right) = 0 \quad (5.24)$$

Equation (5.24) is a quadratic equation, which can be solved for t_h. Solving the quadratic equation for t_h:

$$t_h = \frac{\dfrac{2aT_H}{(a+b)} \pm \sqrt{\left(\dfrac{2aT_H}{(a+b)}\right)^2 - 4\left(\dfrac{a^2 T_H^2 - b^2 T_H T_L}{(a+b)^2}\right)}}{2} \quad (5.25)$$

Squaring and multiplying the expressions under the square root radical of equation (5.25) gives:

$$t_h = \frac{\dfrac{2aT_H}{(a+b)} \pm \sqrt{\dfrac{4a^2 T_H^2}{(a+b)^2} - \dfrac{4a^2 T_H^2}{(a+b)^2} + \dfrac{4b^2 T_H T_L}{(a+b)^2}}}{2} \quad (5.26)$$

The first two terms under the radical cancel out and equation (5.26) becomes:

(5.27)
$$t_h = \frac{\frac{2aT_H}{(a+b)} \pm \frac{2b}{(a+b)}\sqrt{T_H T_L}}{2}$$

Noticing the +/- sign before the second term of the numerator on the right side of the equations indicates two roots to the equation. One root of the equation is:

(5.28)
$$t_h = \frac{aT_H + b\sqrt{T_H T_L}}{(a+b)}$$

The other root of the equation is:

(5.29)
$$t_h = \frac{aT_H - b\sqrt{T_H T_L}}{(a+b)}$$

Now, since the objective of the above derivation is to find the temperature, t_h, when the net heat, H_N, is maximum, it must be determined whether equation (5.28) or equation (5.29) is the correct one to use. Previously, equation (5.7) was derived giving the efficiency of the MEE in terms of t_h and t_l:

$$e = \frac{t_h - t_l}{t_h}$$

When the efficiency, e, is zero, $t_h - t_l$ must also be equal to zero, according to the previous equation. Therefore, in order for e to be zero, t_h must equal t_l. Since t_l had been derived in terms of t_h, giving equation (5.16), the right side of that equation can be set equal to t_h, giving:

$$t_h = \frac{b t_h T_L}{(a+b) t_h - a T_H} \qquad (5.30)$$

Equation (5.30) can be solved for t_h, giving the following equation:

$$t_h = \frac{a T_H + b T_L}{(a+b)} \qquad (5.31)$$

Equation (5.31) gives the value of t_h when the efficiency, e, is zero, when t_h equals t_l, and obviously, when the net work, H_N, is also zero. In order for the net work to be maximum, t_h must be greater than the right side of the equation. So, inspecting equation (5.28) and equation (5.29), it is clear that equation (5.28) is the only equation to give the maximum value of the net heat, H_N. The right side of equation (5.29) cannot be greater than the right side of equation (5.31) since the second term of its numerator is negative. Now, the equation for the maximum net heat can be written by substituting the right side of equation (5.28) for t_h in equation (5.19). Performing the substitution gives the following equation:

$$(5.32) \quad H_N^{Max.} = aT_H - a\left(\frac{aT_H + b\sqrt{T_H T_L}}{(a+b)}\right) - \frac{b^2 T_L\left(\dfrac{aT_H + b\sqrt{T_H T_L}}{(a+b)}\right)}{(a+b)\left(\dfrac{aT_H + b\sqrt{T_H T_L}}{(a+b)}\right) - aT_H} + bT_L$$

The third term on the right side of the equation can be simplified. In the denominator, the (a + b)'s cancel out and the aT_H's cancel out, giving:

56

$$H_N^{Max.} = aT_H - a\left(\frac{aT_H + b\sqrt{T_H T_L}}{a+b}\right) - \frac{b^2 T_L\left(\frac{aT_H + b\sqrt{T_H T_L}}{a+b}\right)}{b\sqrt{T_H T_L}} + bT_L \qquad (5.33)$$

Further simplifying equation (5.33) results in:

$$H_N^{Max.} = aT_H - \frac{a^2 T_H + ab\sqrt{T_H T_L}}{a+b} - \frac{\frac{ab^2 T_H T_L + b^3 T_L\sqrt{T_H T_L}}{a+b}}{b\sqrt{T_H T_L}} + bT_L \qquad (5.34)$$

The third term on the right side can be further simplified by dividing, giving:

$$H_N^{Max.} = aT_H - \frac{a^2 T_H + ab\sqrt{T_H T_L}}{a+b} - \frac{ab\sqrt{T_H T_L} + b^2 T_L}{a+b} + bT_L \qquad (5.35)$$

By multiplying the first and last terms on the right by (a + b) divided by (a + b) gives one fraction on the right:

$$H_N^{Max.} = \frac{a^2 T_H + abT_H - a^2 T_H - 2ab\sqrt{T_H T_L} - b^2 T_L + abT_L + b^2 T_L}{a+b} \qquad (5.36)$$

John D. Jacoby

The first and third and the fifth and seventh terms of the numerator cancel out, giving:

(5.37)
$$H_N^{Max.} = \frac{abT_H - 2ab\sqrt{T_H T_L} + abT_L}{a+b}$$

Equation (5.37) gives the value of the maximum net heat, which is equivalent to the maximum net work. If a, b, T_H, and T_L are known, the value of H_N, maximum, can be calculated.

By dividing the net heat, equation (5.19), by the maximum net heat, equation (5.37), the following equation can be written:

(5.38)
$$\frac{H_N}{H_N^{Max.}} = \frac{aT_H - at_h - \dfrac{b^2 T_L t_h}{(a+b)t_h - aT_H} + bT_L}{\dfrac{abT_H - 2ab\sqrt{T_H T_L} + abT_L}{a+b}}$$

The numerator on the right side of the equation can be simplified and the work relative, W_R, can be substituted for the left side of the equation, since the net work is equivalent to the net

58

heat and the maximum net work is equivalent to the maximum net heat, giving:

$$W_R = \dfrac{\dfrac{2a^2 T_H t_h + ab T_H t_h - a^2 T_H^2 - a^2 t_h^2 - ab t_h^2 + ab T_L t_h - ab T_H T_L}{at_h + bt_h - aT_H}}{\dfrac{ab T_H - 2ab\sqrt{T_H T_L} + ab T_L}{a+b}} \qquad (5.39)$$

Dividing the fraction of the numerator by the fraction of the denominator further simplifies the equation, giving:

$$W_R = \dfrac{(a+b)\left(2a^2 T_H t_h + ab T_H t_h - a^2 T_H^2 - a^2 t_h^2 - ab t_h^2 + ab T_L t_h - ab T_H T_L\right)}{\left(at_h + bt_h - aT_H\right)\left(ab T_H - 2ab\sqrt{T_H T_L} + ab T_L\right)} \qquad (5.40)$$

Now, multiplying both sides of the equation by the denominator, multiplying out the parentheses, and simplifying gives the following equation:

(5.41)
$$\left[(a+b)^2\right]t_h^2$$
$$-\left[a+b\right]\left[(2a+b)T_H+bT_L-bW_R\left(T_H+T_L-2\sqrt{T_HT_L}\right)\right]t_h$$
$$+\left[a\left(a+b-bW_R\right)T_H^2+b\left(a+b-aW_R\right)T_HT_L+2abW_RT_H\sqrt{T_HT_L}\right]=0$$

Equation (5.41) can be recognized as a quadratic equation, which can be solved for t_h. Using equation (5.41) and equation (5.28), graphs can be developed giving values of t_h, the high temperature of the working fluid versus W_R, the net work divided by the maximum net work.

Equation (5.28) gives the temperature, t_h, when the net heat, H_N, is maximum. When H_N is maximum, H_N divided by $H_N^{Max.}$ is equal to one. Since the net heat, H_N, is equivalent to the net work, W_N, and $H_N^{Max.}$ is equivalent to the maximum net work, $W_N^{Max.}$, the relative net work, W_R, is also equal to one at that point.

By inspecting equation (5.28) and equation (5.41), in order to find the various values of t_h, the following variables must be known: a, b, T_H, T_L, and W_R. Assuming that T_H and T_L are known, various values of W_R can be chosen, from 0 to 1. However, without knowing a and b, t_h cannot be determined. But, for purposes of providing some graphs, b can be chosen to be a multiple of a.

For an example of a graph, the following values for the variables will be assumed: b = a, T_H = 1500 R, T_L = 500 R, and values of W_R shall be 0.00, 0.25, 0.50, 0.75, and 1.00.

The first step will be to find the value of t_h, when W_R = 1. That is the value when the work is maximum. Using equation (5.28), the value of t_h when the work is maximum can be determined:

$$t_h = \frac{aT_H + b\sqrt{T_H T_L}}{(a+b)} = \frac{a(1500) + a\sqrt{(1500)(500)}}{2a} = 1183.01$$

So, the value of t_h when the work is maximum is 1183.01 degrees Rankine. That is the value when W_R=1.

61

John D. Jacoby

Now, values of t_h can be determined for the other values of W_R equal to 0.75, 0.50, 0.25, and 0.00 by using the same values, $b=a$, $T_H=1500$, and $T_L=500$, in equation (5.41). With those values of t_h, the graph of t_h versus W_R can be plotted (See Fig. 23, P. 64).

As mentioned, earlier, equation (5.41) is a quadratic equation of the form $Ax^2 + Bx + C = 0$ with the solution being $x=\frac{-B\pm\sqrt{B^2-4AC}}{2A}$, with one root being $x=\frac{-B+\sqrt{B^2-4Ac}}{2A}$ and the other root being $x=\frac{-B-\sqrt{B^2-4AC}}{2A}$.

Now, solving equation (5.41) with $W_R=0.75$, $b=a$, $T_H=1500$, and $T_L=500$, results in:

$$4a^2t_h^2$$
$$-\left[6a^2(1500)+2a^2(500)-1.5a^2\left(1500+500-2\sqrt{1500*500}\right)\right]t_h$$
$$+1.25a^2(1500)^2+1.25a^2(1500*500)+1.5(1500)\sqrt{1500*500}=0$$

Dividing both sides of the equation by a^2 eliminates a^2 from the equation, and performing the arithmetic gives:

62

$$4t_h^2 - 9598.1t_h + 5698557.2 = 0$$

Solving the quadratic equation:

$$t_h = \frac{9598.1 \pm \sqrt{(9598.1)^2 - (4)(4)(5698557.2)}}{(2)(4)}$$

The two values are 1321 and 1078, when $W_R=0.75$.

Using the same process gives 1390 and 1043, when $W_R=0.50$; 1448 and 1019, when $W_R=0.25$; and 1500 and 1000, when $W_R = 0$.

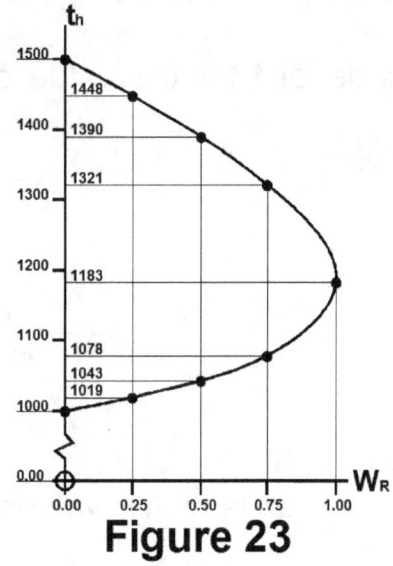

Figure 23

Using those values, the curve is plotted in Figure 23:

The vertical axis is t_h and the horizontal axis is W_R. The vertical axis, t_h, is in degrees Rankine. The horizontal axis units are W_R, the ratio of the net work to the maximum net work. The verti-

63

John D. Jacoby

cal axis is shown broken, since the portion below 1000 is not to scale. As shown, the temperature when the net work is a maximum, when W_R is equal to one, is 1183 degrees Rankine.

Besides having drawn the curve of t_h versus W_R, a curve for t_l versus W_R can also be drawn. Using equation (5.16), for each value of t_h, a value of t_l can be determined. Using that equation, a value of t_l for the value of t_h equal to 1183 at W_R equal to one gives:

$$t_l = \frac{bT_L t_h}{(a+b)t_h - aT_H} = \frac{a(500)1183}{(2a)(1183) - a(1500)} = 683$$

Similarly, the values of t_l for the other values of t_h can be determined. The values of t_l, from high to low, so determined are: 1000, 948, 890, 821, 683, 578, 543, 519, and 500. 1000 degrees Rankine is the value where $t_h = t_l$, and 500 degrees is the value where $t_l = T_L$. Knowing the corresponding values of t_h and t_l, the efficiency of the MEE can be determined for various values. The

64

graph on the facing page shows the curve of t_h versus W_R and the curve of t_l versus W_R.

The graph on the next page, Figure 24, is shown for the parameters, $b = a$, $T_H = 1500$, and $T_L = 500$. Additional graphs on the following pages are for other given values of those parameters.

John D. Jacoby

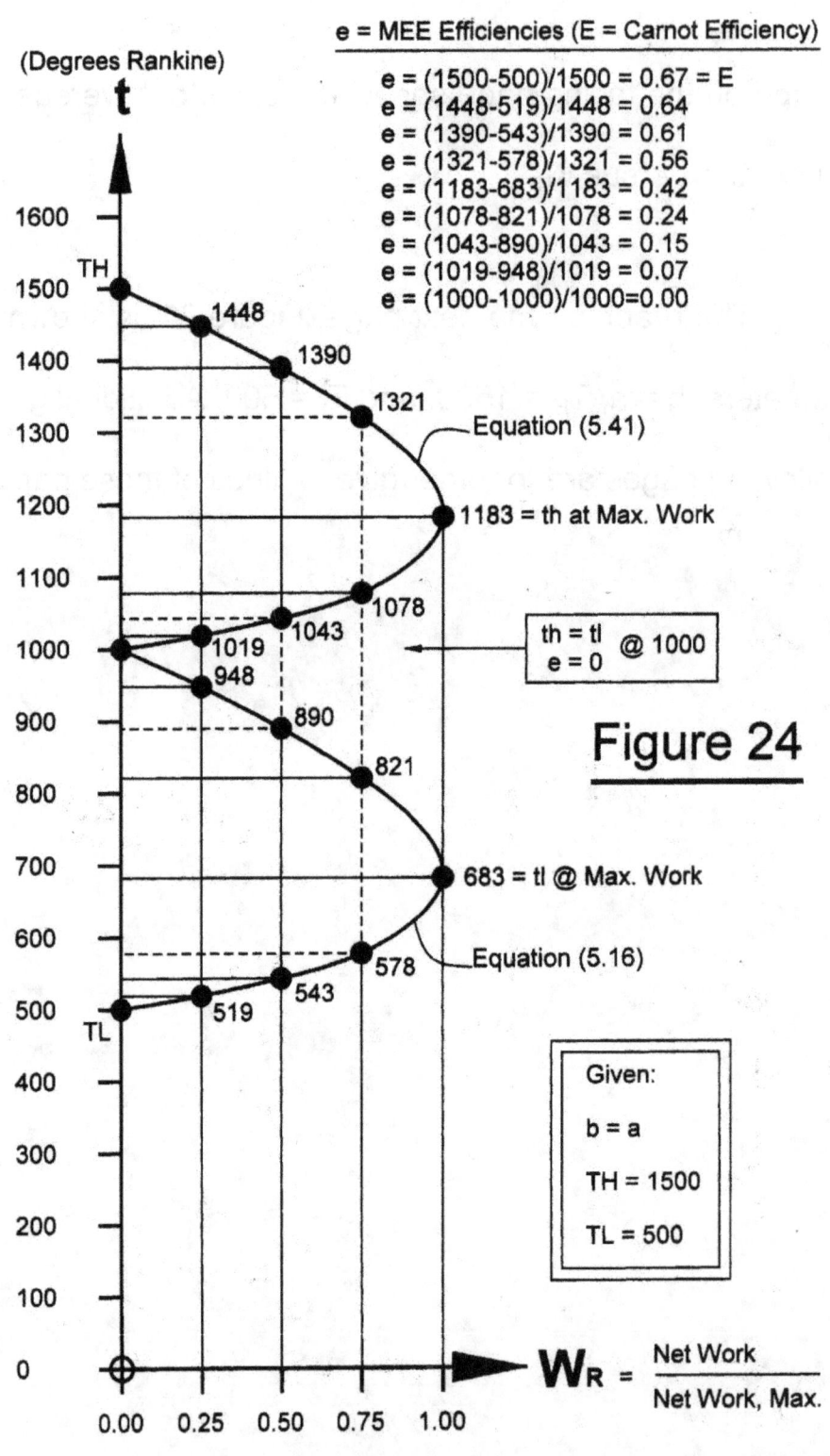

Figure 24

Notice that in Figure 24, t_h temperatures above 1183 correspond to t_l temperatures below 683 and that t_h temperatures below 1183 correspond to t_l temperatures above 683.

The graph of Figure 25 has the given values of b = 1.5a, T_H = 1500, and T_L = 500. It differs from the graph of Figure 24, which has the given values of b = a, T_H =1500, and T_L = 500.

Of course, since the given parameters are different, the temperature values will be different, and the curves will be different. The differences will be obvious when comparing the two graphs. Interestingly, the efficiencies are the same for the graphs of Figure 24 and Figure 25, although the curves are considerably different because of the parameter differences of b = a for Figure 24 and b = 1.5a for Figure 25. For another example, the graphs of Figure 26 and Figure 27 will have different temperature parameters, T_H and T_L.

John D. Jacoby

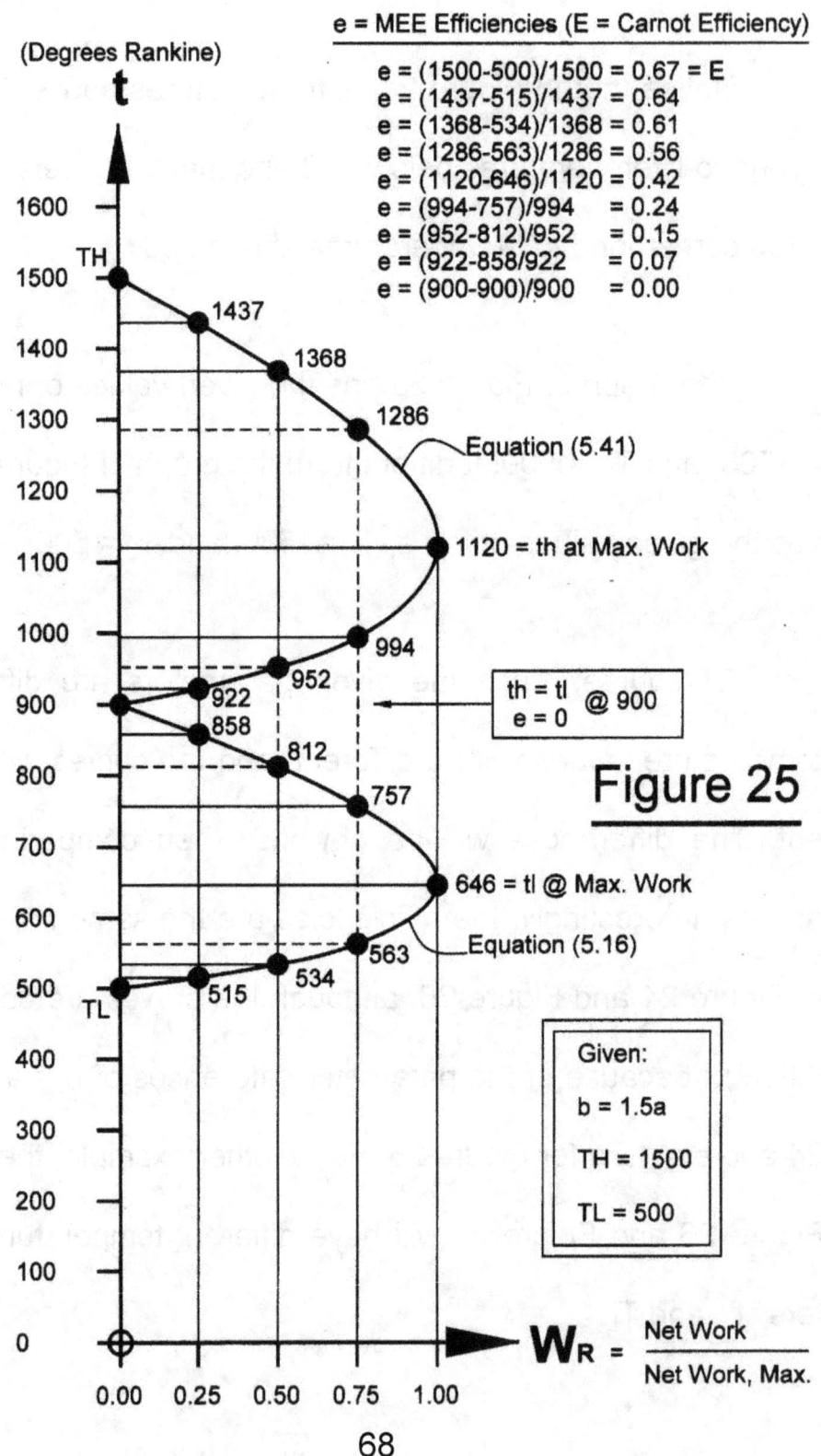

e = MEE Efficiencies (E = Carnot Efficiency)

e = (1500-500)/1500 = 0.67 = E
e = (1437-515)/1437 = 0.64
e = (1368-534)/1368 = 0.61
e = (1286-563)/1286 = 0.56
e = (1120-646)/1120 = 0.42
e = (994-757)/994   = 0.24
e = (952-812)/952   = 0.15
e = (922-858/922   = 0.07
e = (900-900)/900   = 0.00

(Degrees Rankine)

t

1600
1500 TH
1437
1400
1368
1300
1286 — Equation (5.41)
1200
1120 = th at Max. Work
1100
1000
994
952
900   922
      858
800   812
757
700
646 = tl @ Max. Work
600
563 — Equation (5.16)
534
500   515
TL

th = tl @ 900
e = 0

Figure 25

400
300
200
100
0

Given:

b = 1.5a

TH = 1500

TL = 500

$$W_R = \frac{\text{Net Work}}{\text{Net Work, Max.}}$$

0.00 0.25 0.50 0.75 1.00

68

On the next page, the graph of Figure 26 will be similar to the graph of Figure 24, except with $T_H = 660$ and with $T_L = 500$.

The temperature range between T_H and T_L is 160 degrees and is within the temperature range between freezing water at 32 degrees Fahrenheit and boiling water at 212 degrees Fahrenheit, a difference of 180 degrees.

Hopefully, studying these graphs will give a better feeling of the MEE concept than the study of the equations, alone, remembering that the graphs are a result of the equations on the previous pages of this chapter.

John D. Jacoby

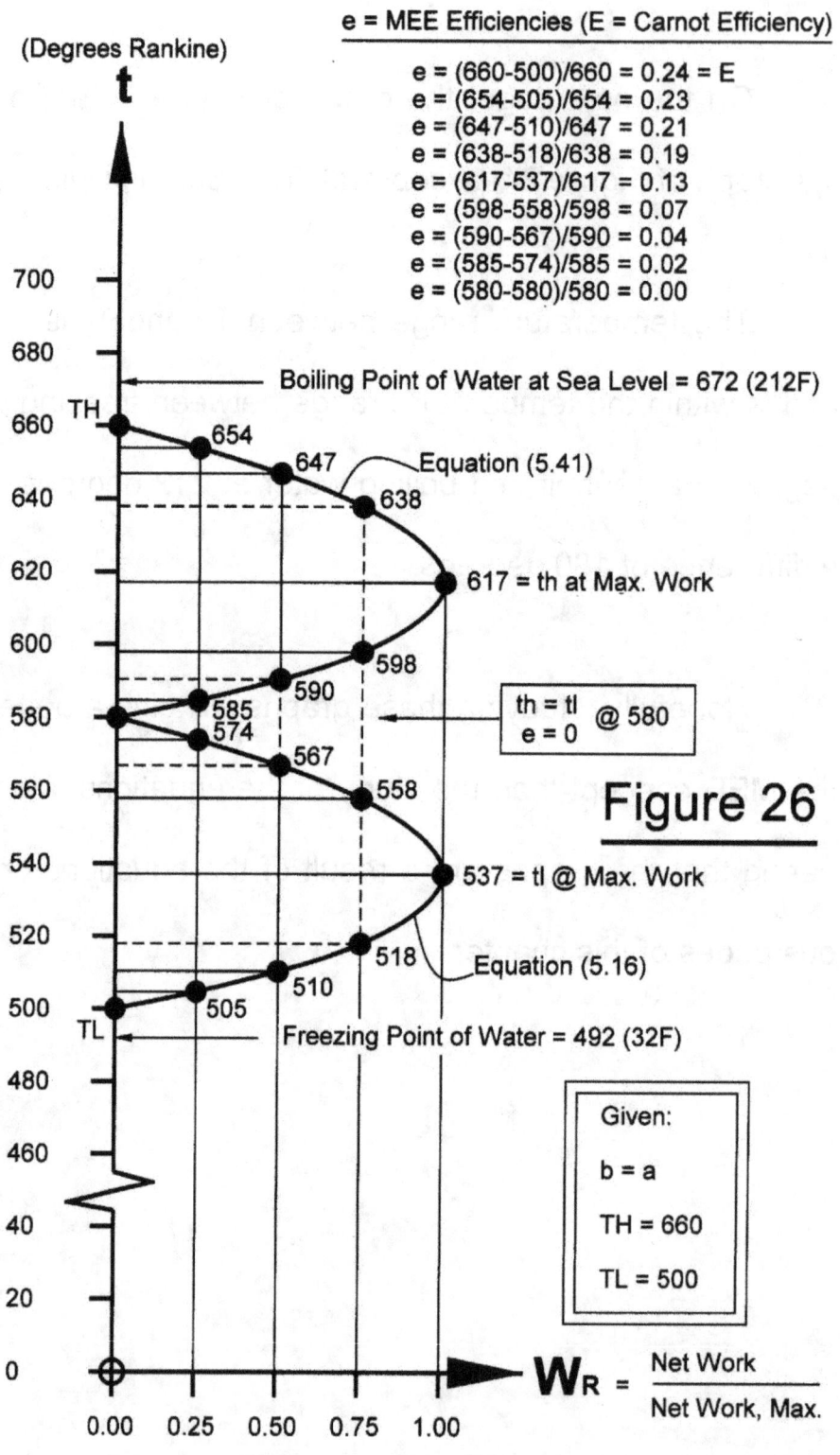

Figure 26

70

Figure 26 on the previous page shows a graph with the given parameters $b = a$, $T_H = 660$, and $T_L = 500$. Figure 27 on the next page is the same, except that $b = 1.5a$. Again, the temperature range is 160 degrees, which is within the temperature range between freezing at 32 degrees Fahrenheit and the boiling point of water at 212 degrees Fahrenheit.

Remembering the list of terms near the beginning of this chapter, a is the heat transfer coefficient at the heat source and b is the heat transfer coefficient at the heat sink. So, as a reminder, assuming that the only difference between a and b are their heat transfer areas, then the heat transfer area at the heat sink is one hundred and fifty percent of the of the heat transfer area at the heat source.

At any rate, Figure 27 is different from Figure 26 because of their ratios of b to a, evidenced by the curve differences.

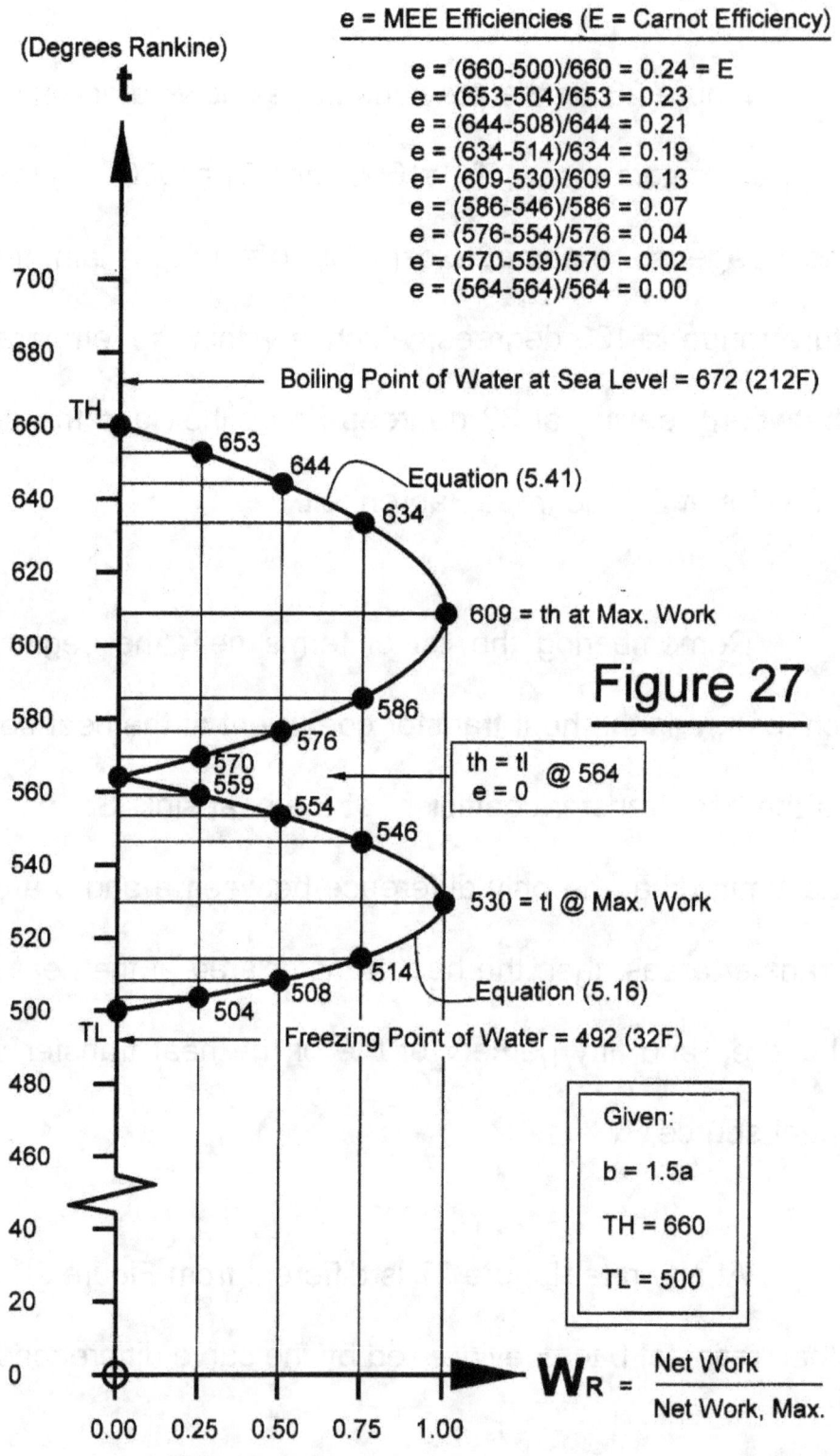

e = MEE Efficiencies (E = Carnot Efficiency)

e = (660-500)/660 = 0.24 = E
e = (653-504)/653 = 0.23
e = (644-508)/644 = 0.21
e = (634-514)/634 = 0.19
e = (609-530)/609 = 0.13
e = (586-546)/586 = 0.07
e = (576-554)/576 = 0.04
e = (570-559)/570 = 0.02
e = (564-564)/564 = 0.00

(Degrees Rankine)
t

Boiling Point of Water at Sea Level = 672 (212F)

TH

653
644   Equation (5.41)
634

609 = th at Max. Work

Figure 27

586
576
570
559   554

th = tl @ 564
e = 0

546

530 = tl @ Max. Work

514
508   Equation (5.16)
504

TL   Freezing Point of Water = 492 (32F)

Given:

b = 1.5a

TH = 660

TL = 500

$W_R = \dfrac{\text{Net Work}}{\text{Net Work, Max.}}$

0.00   0.25   0.50   0.75   1.00

The previous four figures, Figure 24, 25, 26, and 27 are based upon equation (5.41) and equation (5.16). No values for a, nor for b were used. Only the ratios, b = a, and b = 1.5a, were used. However, knowing the values of the heat transfer coefficients, a, and b, would not change the graphs since the ratios could still be determined.

On the following page, Figure 28 is shown. It shows the MEE cycle at five different operating temperature ranges for the case where b = a, T_H = 1500, and T_L = 500. (See Figure 24)

Other such figures, similar to Figure 28, can be portrayed for the many cases with different given values for the heat transfer coefficient ratios and for the temperatures.

Tables in Appendix A provide values for constructing other graphs as desired.

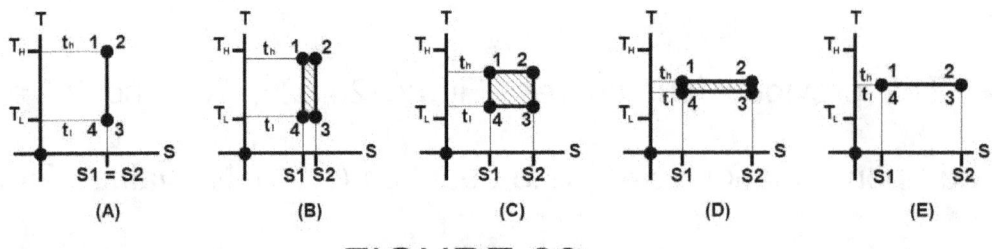

FIGURE 28

Referring to Figure 28, there are five TS diagrams. Each of these diagrams depicts the MEE cycle with different working fluid temperatures, t_h, and t_l. Referring to Figure 24, diagram (A) is the TS diagram for the case where $W_R = 0.00$, $t_h = 1500$, and $t_l = 500$.

Diagram (B) is the case where $W_R = 0.50$, $t_h = 1390$, and $t_l = 543$.

Diagram (C) is the case where $W_R = 1.00$, $t_h = 1183$, and $t_l = 683$.

Diagram (D) is the case where $W_R = 0.50$, $t_h = 1043$, and $t_l = 890$.

Diagram (E) is the case where $W_R = 0.00$, $t_h = 1000$, and $t_l = 1000$.

Diagram (C) is the case where the net work is the maximum. The area bounded by the processes 1-2, 2-3, 3-4, and 4-1 is the largest of the five diagrams. Diagrams (A) and (E) are the cases where the net work is zero since their areas are zero. Cases (B) and (D), are greater than zero and less than the maximum.

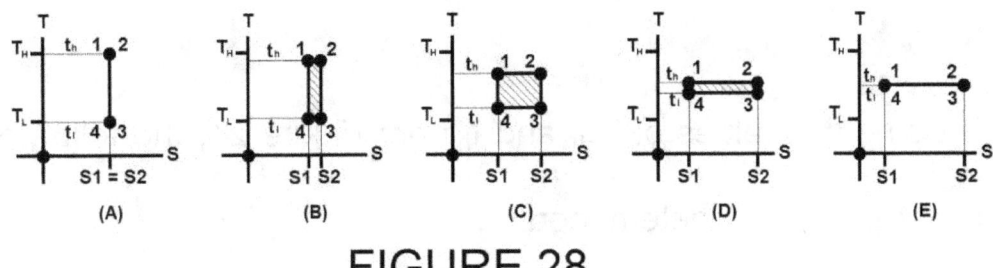

FIGURE 28

Figure 28 is repeated here for ease of reference. The temperatures are shown to scale. The difference in entropies, $(S_2 - S_1)$, in each diagram is not to scale. However, the differences in each diagram are proportional to the others. The $(S_2 - S_1)$ proportions were determined as follows:

From equation (5.1), the heat added is equal to $t_h(S_2 - S_1)$. From equation (5.8), it is also equal to $a(T_H - t_h)$. Therefore, the following equation can be written:

$$t_h(S2 - S1) = a(T_H - t_h) \qquad (5.42)$$

Dividing both sides of the equation by t_h gives the equation for $(S_2 - S_1)$:

75

(5.43)
$$(S2-S1)=\frac{a(T_H-t_h)}{t_h}$$

Using the values of T_H, and t_h from Figure 24, the following values of (s_2-s_1) are determined:

For Figure 28(A), $(s_2-s_1)=0$. For Figure 28(B), $(s_2-s_1)=0.079a$. For Figure 28(C), $(s_2-s_1)=0.268a$. For Figure 28(D), $(s_2-s_1)=0.438a$.

And, for Figure 28(E), $(s_2-s_1)=0.500a$.

So, as can be seen in Figure 28, those proportions of (s_2-s_1) appear to be as stated in the previous paragraph.

Figure 28(A) shows that the isothermal processes 1-2 and 3-4 are non-existent. Figure 28(E) shows that the isentropic processes 2-3 and 4-1 are also non-existent. There are no areas within the processes and, of course, no net work for either one. However, the efficiency for Figure 28(E) is zero and the efficiency for 28(A) is maximum and equal to the Carnot efficiency.

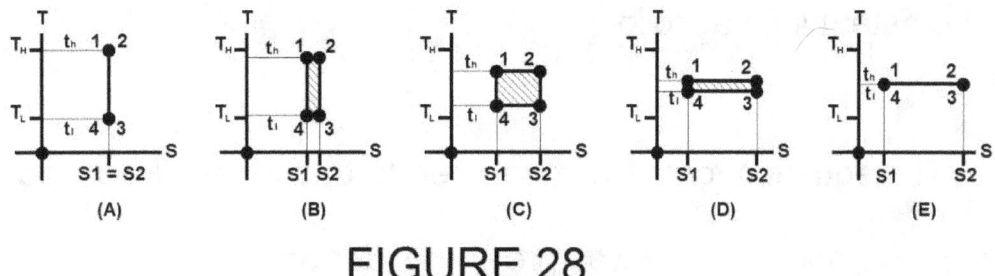

FIGURE 28

Figure 28 is repeated here for convenience. Figure 28(A) (**The New Carnot Cycle**) shows the upper limit of efficiency, equal to the Carnot efficiency, for the MEE engine. The MEE cycle, in this case, consists of only two processes, an isentropic expansion, 2-3, and an isentropic compression, 4-1. This cycle is analogous to a ping pong ball bouncing up and down in a frictionless environment. Similarly, Figure 28(E) consists of only two processes, an isothermal expansion, 1-2, and an isothermal compression, 3-4. The efficiency is zero, since heat is added and the same amount of heat is rejected, no heat being converted to work. This is analogous to a ping pong ball bouncing back and forth horizontally, gathering dust from 1-2 and shedding it from 3-4, in a frictionless environment.

The efficiency of the MEE at maximum net work per cycle can be determined as follows:

The equation for the high temperature of the working fluid at maximum work per cycle was previously derived:

(5.44)
$$t_h = \frac{aT_H + b\sqrt{T_H T_L}}{(a+b)}$$

The equation for the low temperature of the working fluid at maximum work per cycle can be determined from:

(5.45)
$$t_l = \frac{bT_L t_h}{(a+b)t_h - aT_H}$$

Substituting the right side of Equation (5.44) for t_h in equation (5.45) and solving, we get:

(5.46)
$$t_l = \frac{a\sqrt{T_H T_L} + bT_L}{a+b}$$

Now, since efficiency, e = 1 - t_l / t_h, substituting the right

side of equation (5.44) for t_h and the right side of equation (5.46)

for t_l and simplifying, gives:

$$e = 1 - \frac{a\sqrt{T_H T_L} + bT_L}{aT_H + b\sqrt{T_H T_L}} \qquad (5.47)$$

Equation (5.47) gives the efficiency of the MEE, when the

work is maximum.

An interesting, surprising, and amazing sideline is that the

efficiency given by equation (5.47) is equal to the efficiency given

by the equation:

$$e = 1 - \sqrt{T_L / T_H} \qquad (5.48)$$

found on Wikipedia(Wikipedia, the Free Encyclopedia (n.

d.)), for the efficiency at maximum power widely quoted and cred-

ited to Novikov and Chambadal. Using the values of a, b, T_H, and

T_L, where W_R = 1, from the charts of Appendix A, the equations

John D. Jacoby

(5.47) and (5.48) are seen to give <u>exactly</u> the same results. This appears to lend credibility to this book and to equation (5.48). Following are several examples of the exact same results for equations (5.47) and (5.48):

<u>Example 1. a = 1, b = 2, T_H = 3000, T_L = 500, W_R = 1</u>

(5.47) $e = 1 - \dfrac{a\sqrt{T_H T_L} + bT_L}{aT_H + b\sqrt{T_H T_L}}$ = 0.59175 (MEE)

(5.48) $e = 1 - \sqrt{T_L / T_H}$ = 0.59175 (Novikov)

<u>Example 2. a = 1, b = 1.25, T_H = 672, T_L = 492, W_R = 1</u>

(5.47) $e = 1 - \dfrac{a\sqrt{T_H T_L} + bT_L}{aT_H + b\sqrt{T_H T_L}}$ = 0.14434 (MEE)

(5.48) $e = 1 - \sqrt{T_L / T_H}$ = 0.14434 (Novikov)

<u>Example 3. a = 1, b = 1.75, T_H = 2000, T_L = 500, W_R = 1</u>

$$(5.47) \quad e = 1 - \frac{a\sqrt{T_H T_L} + bT_L}{aT_H + b\sqrt{T_H T_L}} = 0.50000 \text{ (MEE)}$$

$$(5.48) \quad e = 1 - \sqrt{T_L / T_H} \quad = 0.50000 \text{ (Novikov)}$$

Those three examples were taken from the charts of Appendix A. It seems like a good validation of those two equations for the efficiency at maximum work per cycle. Try some other examples, referring to the tables of Appendix A.

The graph, shown on the next page, shows efficiency, e, on the X axis, and the relative work per cycle, W_R, on the Y axis. It gives a good picture of the relationship between efficiency and work of Example 1, above. The values for e and W_R are taken from Appendix A.

John D. Jacoby

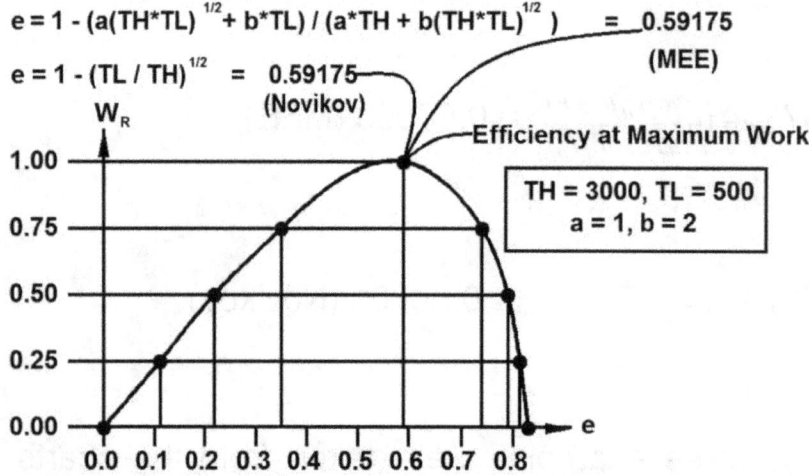

Efficiency vs. Relative Work per Cycle

FIGURE 29

Note that the efficiencies, except for the efficiency at Maximum work are from the MEE cycle only. Other graphs, similar to Figure 29 can be plotted, using the values from Appendix A.

That's all for now, folks! I hope that you enjoyed the trip. And, I hope that the "**Most Efficient Engine**" has been of some value to those of you, who have delved into those crazy equations, figures, and graphs.

82

CHAPTER 6 - SUMMARY

The previous chapters have covered some history, some thermodynamics, the Carnot cycle, the Stirling cycle, and the Most Efficient Engine cycle.

Some thermodynamics has been presented to help with the understanding of *The Most Efficient Engine*. Hopefully, enough fundamentals were presented to allow that. If those fundamentals were carefully studied and understood, then the rest of the book has probably been easier to understand.

With regard to the minimal amount of history presented, it should underline that heat engines are very young in the history of the world. From the beginning of the Industrial Revolution until now has been only about 250 years. Much progress has been made in the development of heat engines. Solar, geothermal, and nuclear energy power heat engines. Automobiles and other vehicles scurry about, ships ply the oceans, jet planes circle the world, and space ships travel to distant planets to study them.

Unfortunately, along with the progress, our advancements also cause air pollution, water pollution, land pollution, and space pollution. More efficient engines will abate these major problems and, hopefully, reduce them to less harmful levels. The MEE is a green engine.

CHAPTER 7 – CONCLUSIONS

Having previously described some thermodynamic fundamentals and various theoretical heat engine cycles, including the Carnot cycle, the Stirling cycle, and the Most Efficient Engine cycle, we can now conclude the following.

The theoretical Carnot cycle still defines the maximum efficiency of all heat engine cycles operating with the same heat source temperature and the same heat sink temperature. In this book, the term *efficiency* denotes *thermal efficiency*. Mechanical efficiency is not included.

The theoretical MEE cycle, developed by the author, also defines the maximum efficiency of all heat engine cycles operating with the same heat source temperature and the same heat sink temperature.

The theoretical Stirling cycle is often described as having the same maximum thermal efficiency. However, in order for that to be true, heat must be added during the constant volume addition of heat at a temperature varying from the low temperature of the working fluid to the high temperature of the working fluid. Heat must be rejected during the constant volume rejection of heat at a temperature varying from the high temperature of the working fluid to the low temperature of the working fluid. Therefore, the Stirling cycle efficiency cannot equal that of the Carnot cycle nor the MEE cycle.

As a practical engine, the Stirling cycle engine is a popular type of heat engine and with the use of regenerators can attain high efficiencies. The regenerators perform the addition and rejection of heat to reuse waste heat to increase efficiency. However, regenerators are not 100 percent efficient and therefore, not all of the rejected heat is recovered.

The Carnot cycle is a theoretical cycle defining the maximum efficiency as described, above. However, an engine designed to operate on the Carnot cycle would be incapable of producing any power at all. The description of the Carnot cycle states that its working fluid receives heat from the heat source at a temperature infinitesimally lower than the heat source temperature and rejects heat to the heat sink at a temperature infinitesimally higher than the heat sink temperature. That means that the heat transfer from the heat source and to the heat sink would occur at an infinitely slow rate. To transfer any meaningful amount of heat would take an infinitely long time. Since power is the rate of performing work, the Carnot cycle would be capable of producing only an infinitesimal amount of power.

The Carnot cycle, in much literature, is often touted as the goal of heat engine designers. Of course, that is a fruitless goal, since a heat engine incapable of producing any power would be meaningless. However, although it is logical and desirable that high efficiency should be a valid goal, high efficiency must be balanced by the goal of producing meaningful power. That would be true of any heat engine.

High efficiency and meaningful power, of course, is desirable for a Stirling cycle engine. Stirling cycle engines, along with other heat engines, must operate with the same goals of high efficiency and meaningful power to be useful.

A heat engine operates at maximum power with efficiency less than its potential efficiency. Conversely, a heat engine can operate with efficiency higher than its efficiency at maximum power, but in order to do so, its power is less than its maximum potential power.

The MEE cycle clearly reveals such efficiency and power relationships. Figure 24, in Chapter 5, is one example of such relationships. For the given parameters, it shows the efficiency at maximum power. As efficiency increases, the power decreases. At the Carnot cycle efficiency, which is equal to the maximum efficiency of the MEE, there can be no power developed. Figure 28 depicts the MEE cycle at various phases, from zero work to maximum work, and from zero efficiency to maximum efficiency equal to the Carnot cycle efficiency. Figure 28 is related to Figure 24, as described in the text of Chapter 5 – *THE MEE CYCLE.*

Since the subject of this book is *THE MOST EFFICIENT ENGINE*, some attention should be paid to it. As mentioned previously, the MEE operates at efficiencies from zero to a maximum efficiency equal to the Carnot cycle efficiency. There must be a temperature differential between the heat source and the working fluid of the engine. There must also be a temperature differential between the working fluid and the heat sink. This is required for heat to flow, just as there must be a difference in elevation in order for water to flow.

Required temperature differentials are clearly demonstrated by referring to Figure 24. When W_R is equal to zero, it is seen that the high temperature of the working fluid equals the temperature of the heat source, 1500 degrees, and the low temperature of the working fluid equals the temperature of the heat sink, 500 degrees. Heat cannot flow and consequently, no work can be performed.

When W_R equals one, work is at its maximum. The high temperature of the working fluid is equal to 1183 degrees and the low temperature of the working fluid is 683 degrees. The temperature differential between the heat source and the working fluid is 317 degrees, allowing considerable heat to flow. The temperature differential between the working fluid and the heat sink is 183 degrees, again allowing considerable heat to flow. With the considerable heat flow, maximum work can be performed.

So, in the case of Figure 24, the temperature difference between the heat source and the heat sink is 1000 degrees. The working fluid traverses a temperature range of only 500 degrees.

Figure 28 depicts the MEE cycle plotted on five separate temperature-entropy diagrams with the working fluid traversing five of the many possible temperature ranges. Since the area within the thermodynamic processes of each diagram indicates net work, a good picture of the variable MEE cycle is presented. Zero work is developed at zero efficiency and zero work is developed at maximum efficiency, with work being produced between those limits.

It is interesting to note that Figure 28(E), depicting the MEE cycle at zero efficiency, shows that there is no change of temperature of the working fluid. The heat received from the heat source is equal to the heat rejected to the heat sink, leaving no heat to perform work. In essence, the working fluid is shown to perform an isothermal expansion followed by an isothermal compression with no isentropic processes. Of course, this cannot happen, since there is no work performed. Figure 28(E) depicts a limit, only. An analogy comes to mind that the working fluid is like a frictionless ping-pong ball bouncing back and forth, right to left and left to right, performing no useful function. Look at Figure 28(A) and see what you come up with. I came up with (The New Carnot Cycle).

CHAPTER 8 - EPILOGUE

Patent # US 8,683,797 B1 has been issued to me for a heat engine based upon the MEE cycle.

The MEE, as I see it, is a heat engine that could operate at maximum power and which could idle at a higher efficiency, with less waste of fuel. It would be a green engine.

John D. Jacoby

Appendix A - Charts and *Appendix B - A Heat Engine* pro-vide some results and a little more understanding of the MEE.

What has been surprising and exciting to me is the fact that I developed the MEE equations with no knowledge of the Novikov and Chambadal equation, which I accidently found on Wikipedia(Wikipedia, the Free Encyclopedia (n. d.)), and mentioned at the end of Chapter 5. And, to learn about it, after the MEE book was almost complete, was extremely gratifying!

Since writing the above paragraph, scratching my head and pondering further, I wondered if the MEE equation for the efficiency at maximum power could be simplified to have exactly the same form as the Novikov equation.

(5.47) $e = 1 - \dfrac{a\sqrt{T_H T_L} + bT_L}{aT_H + b\sqrt{T_H T_L}}$ (MEE)

The second term on the right side of the equation can be simplified as follows:

$$\frac{a\sqrt{T_H T_L} + bT_L}{aT_H + b\sqrt{T_H T_L}} = \frac{a\sqrt{T_H}\sqrt{T_L} + b\sqrt{T_L}\sqrt{T_L}}{a\sqrt{T_H}\sqrt{T_H} + b\sqrt{T_H}\sqrt{T_L}} = \frac{\left(a\sqrt{T_H} + b\sqrt{T_L}\right)\sqrt{T_L}}{\left(a\sqrt{T_H} + b\sqrt{T_L}\right)\sqrt{T_H}} = \sqrt{\frac{T_L}{T_H}}$$

Voila! The simplification gives exactly the same equation as the Novikov equation.

$$e = 1 - \sqrt{\frac{T_L}{T_H}} \qquad \text{(MEE, Novikov)}$$

I wondered why the three examples at the end of Chapter 5 gave the exact same results for the MEE equation and the Novikov equation. I had no idea how the Novikov equation was derived, but that makes it even more gratifying!

Thinking back on the MEE, coming up with the large equation (5.41) would never lead one to believe that the equation for efficiency at maximum power would become such a simple one and one that has also been independently derived by someone else.

Although I have called the equation the Novikov equation, Wikipedia(Wikipedia, the Free Encyclopedia (n. d.)), states that it is often called the Chambadal - Novikov - Curzon - Ahlborn equation, since there seems to be some confusion with regard to the origin of the equation.

Now, on to bigger and better things, as the saying goes!

APPENDIX A – CHARTS

The following pages provide charts with values to allow construction of graphs, such as Figure 24. The second line of each chart displays the given values. The fourth through twelfth lines of each chart display the values needed to plot the graph of the chart values. The charts were developed using the equations in this book and using Microsoft Excel, a wonderful tool. Enough charts have been included to give a broad range of given values.

The value of t_h, when $W_R = 1.00$, is determined using equation (5.28). The values of t_h corresponding to the other values of W_R in the chart are determined by solving equation (5.41), giving two roots to the equation. The root including the positive square root applies to the values above $W_R = 1.00$. The root including the negative square root applies to the values below $W_R = 1.00$. The values of t_l are determined using equation (5.16). The values of e are determined using equation (5.7).

The values in the charts have been determined using the following given values:

b/a equal to 1.00, 1.25, 1.50, 1.75, and 2.00.

T_H equal to 3000, 2500, 2000, 1500, 1000, 660, and 672.

T_L equal to 500 and 492.

There are a total of 35 charts on the following pages.

Appendix A - Charts

b\a	TH	TL	
1.00	3000.00	500.00	

th	WR	tl	e
3000.00	0.00	500.00	0.83
2838.53	0.25	530.16	0.81
2665.67	0.50	571.70	0.79
2469.10	0.75	636.96	0.74
2112.37	1.00	862.37	0.59
1886.96	0.75	1219.10	0.35
1821.70	0.50	1415.67	0.22
1780.16	0.25	1588.53	0.11
1750.00	0.00	1750.00	0.00

b\a	TH	TL	
1.25	3000.00	500.00	

th	WR	tl	e
3000.00	0.00	500.00	0.83
2820.59	0.25	526.81	0.81
2628.52	0.50	563.74	0.79
2410.11	0.75	621.74	0.74
2013.75	1.00	822.11	0.59
1763.28	0.75	1139.20	0.35
1690.78	0.50	1313.93	0.22
1644.62	0.25	1467.58	0.11
1611.11	0.00	1611.11	0.00

b\a	TH	TL	
1.50	3000.00	500.00	
th	WR	tl	e
3000.00	0.00	500.00	0.83
2806.23	0.25	524.13	0.81
2598.80	0.50	557.36	0.79
2362.92	0.75	609.56	0.74
1934.85	1.00	789.90	0.59
1664.35	0.75	1075.28	0.35
1586.04	0.50	1232.53	0.22
1536.19	0.25	1370.82	0.11
1500.00	0.00	1500.00	0.00

b\a	TH	TL	
1.75	3000.00	500.00	
th	WR	tl	e
3000.00	0.00	500.00	0.83
2794.49	0.25	521.93	0.81
2574.49	0.50	552.15	0.79
2324.31	0.75	599.60	0.74
1870.29	1.00	763.54	0.59
1583.40	0.75	1022.98	0.35
1500.35	0.50	1165.94	0.22
1447.47	0.25	1291.66	0.11
1409.09	0.00	1409.09	0.00

Appendix A - Charts

b\a	TH	TL
2.00	3000.00	500.00

th	WR	tl	e
3000.00	0.00	500.00	0.83
2784.70	0.25	520.11	0.81
2554.22	0.50	547.80	0.79
2292.14	0.75	591.30	0.74
1816.50	1.00	741.58	0.59
1515.94	0.75	979.40	0.35
1428.94	0.50	1110.45	0.22
1373.54	0.25	1225.69	0.11
1333.33	0.00	1333.33	0.00

b\a	TH	TL
1.00	2500.00	500.00

th	WR	tl	e
2500.00	0.00	500.00	0.80
2377.30	0.25	527.21	0.78
2244.92	0.50	564.09	0.75
2092.69	0.75	620.84	0.70
1809.02	1.00	809.02	0.55
1620.84	0.75	1092.69	0.33
1564.09	0.50	1244.92	0.20
1527.21	0.25	1377.30	0.10
1500.00	0.00	1500.00	0.00

b\a	TH	TL	
1.25	2500.00	500.00	
th	WR	tl	e
2500.00	0.00	500.00	0.80
2363.66	0.25	524.19	0.78
2216.58	0.50	556.97	0.75
2047.43	0.75	607.41	0.70
1732.24	1.00	774.68	0.55
1523.15	0.75	1026.83	0.33
1460.11	0.50	1162.15	0.20
1419.12	0.25	1279.82	0.10
1388.89	0.00	1388.89	0.00

b\a	TH	TL	
1.50	2500.00	500.00	
th	WR	tl	e
2500.00	0.00	500.00	0.80
2352.76	0.25	521.77	0.78
2193.91	0.50	551.28	0.75
2011.23	0.75	596.67	0.70
1670.82	1.00	747.21	0.55
1445.00	0.75	974.15	0.33
1376.91	0.50	1095.94	0.20
1332.65	0.25	1201.84	0.10
1300.00	0.00	1300.00	0.00

Appendix A - Charts

b\a	TH	TL
1.75	2500.00	500.00

th	WR	tl	e
2500.00	0.00	500.00	0.80
2343.83	0.25	519.79	0.78
2175.36	0.50	546.61	0.75
1981.60	0.75	587.88	0.70
1620.57	1.00	724.74	0.55
1381.07	0.75	931.05	0.33
1308.85	0.50	1041.76	0.20
1261.91	0.25	1138.03	0.10
1227.27	0.00	1227.27	0.00

b\a	TH	TL
2.00	2500.00	500.00

th	WR	tl	e
2500.00	0.00	500.00	0.80
2336.40	0.25	518.14	0.78
2159.90	0.50	542.73	0.75
1956.92	0.75	580.56	0.70
1578.69	1.00	706.01	0.55
1327.78	0.75	895.13	0.33
1252.13	0.50	996.61	0.20
1202.95	0.25	1084.86	0.10
1166.67	0.00	1166.67	0.00

b\a	TH	TL	
1.00	2000.00	500.00	

th	WR	tl	e
2000.00	0.00	500.00	0.75
1913.97	0.25	523.53	0.73
1820.19	0.50	554.81	0.70
1710.77	0.75	601.73	0.65
1500.00	1.00	750.00	0.50
1351.73	0.75	960.77	0.29
1304.81	0.50	1070.19	0.18
1273.53	0.25	1163.97	0.09
1250.00	0.00	1250.00	0.00

b\a	TH	TL	
1.25	2000.00	500.00	

th	WR	tl	e
2000.00	0.00	500.00	0.75
1904.41	0.25	520.92	0.73
1800.22	0.50	548.72	0.70
1678.63	0.75	590.43	0.65
1444.44	1.00	722.22	0.50
1279.70	0.75	909.57	0.29
1227.56	0.50	1006.84	0.18
1192.81	0.25	1090.19	0.09
1166.67	0.00	1166.67	0.00

Appendix A - Charts

b\a	TH	TL	
1.50	2000.00	500.00	

th	WR	tl	e
2000.00	0.00	500.00	0.75
1896.76	0.25	518.83	0.73
1784.23	0.50	543.84	0.70
1652.92	0.75	581.39	0.65
1400.00	1.00	700.00	0.50
1222.08	0.75	868.61	0.29
1165.77	0.50	956.16	0.18
1128.24	0.25	1031.17	0.09
1100.00	0.00	1100.00	0.00

b\a	TH	TL	
1.75	2000.00	500.00	

th	WR	tl	e
2000.00	0.00	500.00	0.75
1890.50	0.25	517.11	0.73
1771.16	0.50	539.86	0.70
1631.89	0.75	573.99	0.65
1363.64	1.00	681.82	0.50
1174.93	0.75	835.10	0.29
1115.21	0.50	914.69	0.18
1075.41	0.25	982.89	0.09
1045.45	0.00	1045.45	0.00

b\a	TH	TL	
2.00	2000.00	500.00	

th	WR	tl	e
2000.00	0.00	500.00	0.75
1885.29	0.25	515.69	0.73
1760.26	0.50	536.54	0.70
1614.36	0.75	567.82	0.65
1333.33	1.00	666.67	0.50
1135.64	0.75	807.18	0.29
1073.07	0.50	880.13	0.18
1031.38	0.25	942.64	0.09
1000.00	0.00	1000.00	0.00

b\a	TH	TL	
1.00	1500.00	500.00	

th	WR	tl	e
1500.00	0.00	500.00	0.67
1447.81	0.25	518.70	0.64
1390.08	0.50	542.93	0.61
1321.35	0.75	578.17	0.56
1183.01	1.00	683.01	0.42
1078.17	0.75	821.35	0.24
1042.93	0.50	890.08	0.15
1018.70	0.25	947.81	0.07
1000.00	0.00	1000.00	0.00

Appendix A - Charts

b\a	TH	TL
1.25	1500.00	500.00

th	WR	tl	e
1500.00	0.00	500.00	0.67
1442.01	0.25	516.62	0.64
1377.87	0.50	538.16	0.61
1301.50	0.75	569.49	0.56
1147.79	1.00	662.68	0.42
1031.30	0.75	785.64	0.24
992.15	0.50	846.74	0.15
965.22	0.25	898.05	0.07
944.44	0.00	944.44	0.00

b\a	TH	TL
1.50	1500.00	500.00

th	WR	tl	e
1500.00	0.00	500.00	0.67
1437.37	0.25	514.96	0.64
1368.10	0.50	534.35	0.61
1285.62	0.75	562.54	0.56
1119.62	1.00	646.41	0.42
993.81	0.75	757.08	0.24
951.52	0.50	812.06	0.15
922.44	0.25	858.25	0.07
900.00	0.00	900.00	0.00

b\a	TH	TL	
1.75	1500.00	500.00	

th	WR	tl	e
1500.00	0.00	500.00	0.67
1433.57	0.25	513.60	0.64
1360.10	0.50	531.22	0.61
1272.62	0.75	556.85	0.56
1096.56	1.00	633.10	0.42
963.13	0.75	733.71	0.24
918.28	0.50	783.70	0.15
887.43	0.25	825.68	0.07
863.64	0.00	863.64	0.00

b\a	TH	TL	
2.00	1500.00	500.00	

th	WR	tl	e
1500.00	0.00	500.00	0.67
1430.41	0.25	512.47	0.64
1353.44	0.50	528.62	0.61
1261.80	0.75	552.11	0.56
1077.35	1.00	622.01	0.42
937.56	0.75	714.23	0.24
890.58	0.50	760.05	0.15
858.26	0.25	798.54	0.07
833.33	0.00	833.33	0.00

Appendix A - Charts

b\a	TH	TL
1.00	1000.00	500.00

th	WR	tl	e
1000.00	0.00	500.00	0.50
977.49	0.25	511.78	0.48
952.01	0.50	526.54	0.45
920.72	0.75	547.11	0.41
853.55	1.00	603.55	0.29
797.11	0.75	670.72	0.16
776.54	0.50	702.01	0.10
761.78	0.25	727.49	0.05
750.00	0.00	750.00	0.00

b\a	TH	TL
1.25	1000.00	500.00

th	WR	tl	e
1000.00	0.00	500.00	0.50
974.99	0.25	510.47	0.48
946.68	0.50	523.59	0.45
911.91	0.75	541.87	0.41
837.28	1.00	592.05	0.29
774.57	0.75	651.75	0.16
751.71	0.50	679.57	0.10
735.32	0.25	702.22	0.05
722.22	0.00	722.22	0.00

John D. Jacoby

b\a	TH	TL	
1.50	1000.00	500.00	

th	WR	tl	e
1000.00	0.00	500.00	0.50
972.99	0.25	509.43	0.48
942.41	0.50	521.23	0.45
904.87	0.75	537.69	0.41
824.26	1.00	582.84	0.29
756.53	0.75	636.58	0.16
731.85	0.50	661.61	0.10
714.14	0.25	681.99	0.05
700.00	0.00	700.00	0.00

b\a	TH	TL	
1.75	1000.00	500.00	

th	WR	tl	e
1000.00	0.00	500.00	0.50
971.35	0.25	508.57	0.48
938.92	0.50	519.30	0.45
899.10	0.75	534.26	0.41
813.61	1.00	575.31	0.29
741.77	0.75	624.16	0.16
715.60	0.50	646.92	0.10
696.82	0.25	665.45	0.05
681.82	0.00	681.82	0.00

Appendix A - Charts

b\a	TH	TL
2.00	1000.00	500.00

th	WR	tl	e
1000.00	0.00	500.00	0.50
969.99	0.25	507.86	0.48
936.02	0.50	517.69	0.45
894.30	0.75	531.41	0.41
804.74	1.00	569.04	0.29
729.48	0.75	613.81	0.16
702.05	0.50	634.67	0.10
682.38	0.25	651.66	0.05
666.67	0.00	666.67	0.00

b\a	TH	TL
1.00	660.00	500.00

th	WR	tl	e
660.00	0.00	500.00	0.24
653.93	0.25	504.69	0.23
646.86	0.50	510.36	0.21
637.89	0.75	517.96	0.19
617.23	1.00	537.23	0.13
597.96	0.75	557.89	0.07
590.36	0.50	566.86	0.04
584.69	0.25	573.93	0.02
580.00	0.00	580.00	0.00

b\a	TH	TL	
1.25	660.00	500.00	

th	WR	tl	e
660.00	0.00	500.00	0.24
653.25	0.25	504.17	0.23
645.40	0.50	509.21	0.21
635.43	0.75	515.96	0.19
612.48	1.00	533.09	0.13
591.06	0.75	551.45	0.07
582.63	0.50	559.43	0.04
576.32	0.25	565.71	0.02
571.11	0.00	571.11	0.00

b\a	TH	TL	
1.50	660.00	500.00	

th	WR	tl	e
660.00	0.00	500.00	0.24
652.71	0.25	503.75	0.23
644.24	0.50	508.29	0.21
633.46	0.75	514.37	0.19
608.67	1.00	529.78	0.13
585.55	0.75	546.31	0.07
576.44	0.50	553.49	0.04
569.62	0.25	559.14	0.02
564.00	0.00	564.00	0.00

Appendix A - Charts

b\a	TH	TL
1.75	660.00	500.00

th	WR	tl	e
660.00	0.00	500.00	0.24
652.27	0.25	503.41	0.23
643.28	0.50	507.54	0.21
631.85	0.75	513.06	0.19
605.56	1.00	527.08	0.13
581.04	0.75	542.10	0.07
571.37	0.50	548.63	0.04
564.15	0.25	553.77	0.02
558.18	0.00	558.18	0.00

b\a	TH	TL
2.00	660.00	500.00

th	WR	tl	e
660.00	0.00	500.00	0.24
651.90	0.25	503.12	0.23
642.49	0.50	506.91	0.21
630.51	0.75	511.97	0.19
602.97	1.00	524.82	0.13
577.28	0.75	538.59	0.07
567.15	0.50	544.58	0.04
559.58	0.25	549.28	0.02
553.33	0.00	553.33	0.00

b\a	TH	TL	
1.00	672.00	492.00	

th	WR	tl	e
672.00	0.00	492.00	0.27
665.07	0.25	497.18	0.25
657.02	0.50	503.48	0.23
646.82	0.75	511.93	0.21
623.50	1.00	533.50	0.14
601.93	0.75	556.82	0.07
593.48	0.50	567.02	0.04
587.18	0.25	575.07	0.02
582.00	0.00	582.00	0.00

b\a	TH	TL	
1.25	672.00	492.00	

th	WR	tl	e
672.00	0.00	492.00	0.27
664.30	0.25	496.61	0.25
655.36	0.50	502.20	0.23
644.03	0.75	509.71	0.21
618.11	1.00	528.89	0.14
594.14	0.75	549.62	0.07
584.75	0.50	558.69	0.04
577.76	0.25	565.84	0.02
572.00	0.00	572.00	0.00

Appendix A - Charts

b\a	TH	TL	
1.50	672.00	492.00	

th	WR	tl	e
672.00	0.00	492.00	0.27
663.68	0.25	496.15	0.25
654.03	0.50	501.18	0.23
641.79	0.75	507.94	0.21
613.80	1.00	525.20	0.14
587.91	0.75	543.86	0.07
577.77	0.50	552.02	0.04
570.22	0.25	558.45	0.02
564.00	0.00	564.00	0.00

b\a	TH	TL	
1.75	672.00	492.00	

th	WR	tl	e
672.00	0.00	492.00	0.27
663.18	0.25	495.77	0.25
652.94	0.50	500.35	0.23
639.96	0.75	506.49	0.21
610.27	1.00	522.18	0.14
582.81	0.75	539.14	0.07
572.06	0.50	546.56	0.04
564.05	0.25	552.41	0.02
557.45	0.00	557.45	0.00

b/a	TH	TL	
2.00	672.00	492.00	

th	Wr	tl	e
672.00	0.00	492.00	0.27
662.76	0.25	495.46	0.25
652.03	0.50	499.65	0.23
638.43	0.75	505.28	0.21
607.33	1.00	519.67	0.14
578.57	0.75	535.22	0.07
567.30	0.50	542.01	0.04
558.91	0.25	547.38	0.02
552.00	0.00	552.00	0.00

In case the above table seems out of place, the reason it is there is because 672 degrees Rankine equals 212 degrees Fahrenheit and 492 degrees Rankine equals 32 degrees Fahrenheit.

212 degrees Fahrenheit is the boiling point of water at Sea Level and 32 degrees Fahrenheit is the freezing point of water, a range of 180 degrees Fahrenheit.

APPENDIX B - A HEAT ENGINE

The following pages depict two very simplified heat engines, based on my patent # US 8,683,797 B1.

Both of them are rotary engines. Engine 1 is a rotary vane type engine and engine 2 is a rotary piston type engine. They are two examples of heat engines, which can be designed to operate on the MEE cycle. Perhaps that will allow you to better visualize the relevance of the MEE cycle to real heat engines.

The terms in parentheses are related to the MEE cycle analyzed in Chapter 5.

John D. Jacoby

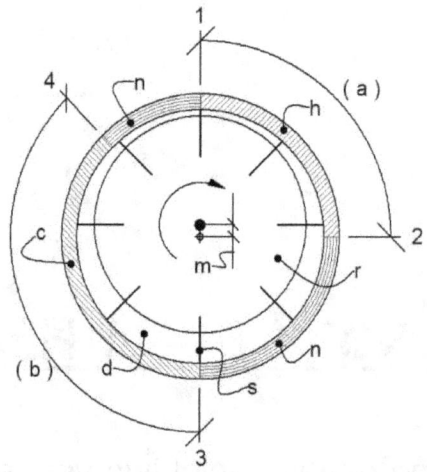

<u>Engine 1 - Rotary Vane Type</u>

h = Hot Zone at Temperature, (T_H).
c = Cold Zone at Temperature, (T_L).
n = Neutral Zone with low thermal conductivity.
r = Rotor mounted eccentrically.
s = Vane
m = eccentricity of Rotor, r
d = Variable volume working chamber with confined working fluid.

The working fluid is at temperature (t_h) from 1 to 2, varies from (t_h) to (t_l) from 2 to 3, is at temperature (t_l) from 3 to 4, and varies from (t_l) to (t_h) from 4 to 1.

118

(a) is the length of the Hot Zone, h, and (b) is the length of the Cold Zone, c.

For Engine 1, the Hot Zone and Cold Zone properties are assumed to have the same heat transfer properties, except for their lengths. Therefore, the ratio of (b) to (a) is proportional to their lengths.

For operation, the rotor diameter, the eccentricity, and the lengths (a) and (b) are manipulated to meet the predetermined MEE requirements. (a) begins at the minimum volume of the working chambers and (b) begins at the maximum volume of the working chambers. The rotor rotates in the clockwise direction. The volume of each working chamber at 2 divided by the volume of each working chamber at 3 should be equal to the volume of each working chamber at 1 divided by the volume of each working chamber at 4 to assure that the isentropic expansion is equal to the isentropic compression and the temperatures, (t_h) and (t_l) are maintained.

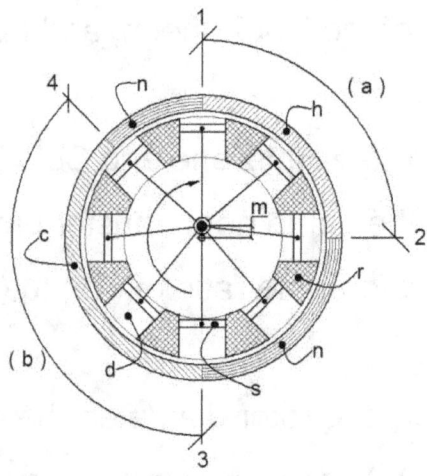

Engine 2 - Piston Type

h = Hot Zone at Temperature, (T_H).
c = Cold Zone at Temperature, (T_L).
n = Neutral Zone with low thermal conductivity.
r = Rotor
s = Piston
m = Eccentric connecting rod post
d = Variable volume working chamber with confined working
fluid.

The working fluid is at temperature (t_h) from 1 to 2, varies from (t_h) to (tl) from 2 to 3, is at temperature (t_l)from 3 to 4, and varies from (t_l) to (t_h) from 4 to 1.

(a) is the length of the Hot Zone, h, and(b) is the length of the Cold Zone, c.

For Engine 2, the Hot Zone and Cold Zone properties are assumed to have the same heat transfer properties, except for their lengths. Therefore, the ratio of (b) to (a) is proportional to their lengths.

For operation, the eccentricity, the connecting rod lengths, and the lengths (a) and (b) are manipulated to meet the predetermined MEE requirements.   (a) begins at the minimum volume of the working chambers and (b) begins at the maximum volume of the working chambers. The rotor rotates in the clockwise direction. The volume of each working chamber at 2 divided by the volume of each working chamber at 3 should be equal to the volume of each working chamber at 1 divided by the volume of each working chamber at 4 to assure that the isentropic expansion is equal to the isentropic compression and that the temperatures, (t_h) and (t_l) are maintained.

BIBLIOGRAPHY

De Laval Turbine Inc. *De Laval Engineering Handbook.* New York: McGraw-Hill Book Company, 1970.

Anderson, Herbert L. *A Physicist's Desk Reference.* New York: American Institute of Physics, 1989.

Atkins, Peter. *The Laws of Thermodynamics.* New York: Oxford University Press, 2010.

Barnard, William N., Frank O. Ellenwood, and Clarence F. Hirshfeld. *Heat-Power Engineering.* New York: John Wiley & Sons, Inc., 1935.

Carnot, Sadi. *Reflexions sur la Puissance Motrice Du Feu et Sur Les Machines Propres a Developper Cette puissance.* Paris: Carnot, 1824.

John D. Jacoby

Daniels, Farrington. *Direct Use of the Sun's Energy.* New Haven and London: Yale University Press, 1964.

Glasstone, Samuel. *Energy Deskbook.* Oak Ridge Tennessee: Technical Information Center, United States Department of Energy, 1982.

Hildebrand, Joel H. *An Introduction to Molecular Kinetic Theory.* New York: Reinhold Publishing Corporation, 1963.

Hudson, Alvin and Rex Nelson. *University Physics.* New York: Harcourt Brace Jovanovich, Inc., 1982.

Hughes, William F. and Eber W. Gaylord. *Basic Eauations of Engineering Science.* New York: McGraw-Hill Book Company (Schaum's Outline Series), 1964.

Jeans, Sir James. *An Introduction to the Kinetic Theory of Gases.* Cambridge: Cambridge University Press, 1962.

Bibliography

Keenan, Joseph H. and Joseph Kaye. *Thermodynamic Properties of Air.* New York: John Wiley & Sons, Inc., 1947.

Kreith, Frank. *Principles of Heat Transfer.* New York: IEP - A Dun-Donnelley Publisher, 1973.

Mendoza, E., R. H. Thurston, and W. F. Magie. *(Trans.) Reflections on the Motive Power of Fire.* Glouster, Mass.: Peter Smith, 1977.

Pitts, Donald R. and Leighton E. Sissom. *Theory and Problems of Heat Transfer.* New York: McGraw-Hill Book Company (Schaum's Outline Series), 1977.

Raznjevic, Kuzman. *Handbook of Thermodynamic Tables and Charts.* Washington and London: Hemisphere Publishing Corporation, 1976.

Reynolds, William C. and Henry C. Perkins. *Engineering Thermodynamics.* New York: McGraw-Hill Book Company, 1977.

Rohsenow, Warren M. and James P. Hartnet. *Handbook of Heat Transfer.* New York: McGraw-Hill Book Company, 1973.

Sandfort, John F. *Heat Engines.* Garden City, New York: Doubleday & Company, Inc., 1962.

Spielberg, Nathan and Bryon D. Anderson. *Seven Ideas that Shook the Universe.* New York: John Wiley and Sons, Inc., 1987.

Tipler, Paul A. *Physics.* New York: Worth Publishers, Inc., 1976.

Van Arsdell, Brent H. *Around the World by Stirling Engine.* San Diego, California: American Stirling Company, 2003.

Bibliography

Van Ness, H. C. and M. M. Abbott. *Theory and Problems of Thermodynamics.* Schaum's Outline Series, New York: McGraw-Hill Book Company, 1972.

Van Ness, H. C. *Understanding Thermodynamics.* New York: Dover Publications, Inc., 1983.

Waldram, J. R. *The Theory of Thermodynamics.* Cambridge: Cambridge University Press, 1987.

Walker, Graham. *Stirling-Cycle Machines.* Oxford: Clarendon Press, 1973.

Warner, Cecil F. *Thermodynamic Fundamentals for Engineers.* Paterson, New Jersey: Littlefield, Adams & Co., 1964.

West, C. D. *Liquid Piston Stirling Engines.* New York: Van Nostrand Reinhold Company, 1983.

John D. Jacoby

Wikipedia, the Free Encyclopedia. *Endoreversible Thermodynamics.* Encyclopedia, internet: Wikimedia Foundation, Inc., (n. d.).

<u>INDEX</u>

129

Index

1953   I hope you enjoyed it! Remember, I welcome any comments to make this a better book - Sincerely, *John D. Jacoby*

JOHNDJACOBY@HOTMAIL.COM